The INVENTION *of* IMAGINATION

The INVENTION *of* IMAGINATION

Aristotle, Geometry, *and the* Theory *of the* Psyche

JUSTIN HUMPHREYS

University of Pittsburgh Press

Published by the University of Pittsburgh Press, Pittsburgh, Pa., 15260

Manufactured in the United States of America
Printed on acid-free paper
10 9 8 7 6 5 4 3 2 1

Cataloging-in-Publication data is available from the Library of Congress

ISBN 13: 978-0-8229-4740-0
ISBN 10: 0-8229-4740-4

Cover art: Abraham Bosse, plate 3 from *Manière universelle de M. Desargues pour pratiquer la perspective par petit pied, comme le géométral* (Paris: Des-Hayes, 1648).

Cover design: Alex Wolfe

In memory of my father,
David Harding Humphreys

ὡς ἡ μὲν ὅρασις ἀπὸ τοῦ περιέχοντος ἀέρος
λαμβάνει τὸ φῶς, ἡ δὲ ψυχὴ ἀπὸ τῶν μαθημάτων.

CONTENTS

Acknowledgments
xi
A Note on Greek Text
xiii
Prologue
3
CHAPTER 1
Images *and* Appearances
9
CHAPTER 2
Powers *of the* Psyche
27
CHAPTER 3
Imagination Defined
40
CHAPTER 4
Dreaming *and* Daydreaming
49

CHAPTER 5

Abstracting

63

CHAPTER 6

Geometrical Thinking

78

CHAPTER 7

Wishing

88

CHAPTER 8

Deliberating

99

CHAPTER 9

Shame *and the* Other

110

Epilogue

124

List of Abbreviations

131

Notes

133

Bibliography

145

Index

153

ACKNOWLEDGMENTS

I thank Dmitri Nikulin, Ronald Polansky, Cinzia Arruzza, and Michael Weinman for offering encouragement and valuable comments on earlier versions of this book. Gisela Striker, Joseph Lemelin, and Jonathan Buttaci read and corrected individual chapters. Discussions with Marcelo Boeri, Duane Lacey, and Darren Gardner allowed me to see ordinary ascriptions of imagination as extraordinary, and hence as calling for investigation. Several colleagues at Villanova, especially Delia Popa, Julie Klein, and John Carvahlo, pushed me to make my interpretations more precise. Finally, I thank Maria Cuellar Correa for more than I can say, but not least for drawing the diagrams.

A NOTE ON GREEK TEXT

This book uses Greek text when it is necessary for grasping an argument. In most cases, this requires citing the Greek when a term is introduced and using its English equivalent thereafter. In the case of a few common words, for which the use of translations could be a source of confusion, Greek terms are transliterated in Latin letters. Most importantly, "phantasia" is used to indicate the general term for appearance, "imagination" being employed when the word is meant in its marked sense as a psychological power. A state produced by this power is a "phantasma," the plural of which is "phantasmata."

The INVENTION *of*
IMAGINATION

PROLOGUE

How are synthetic judgments *a priori possible*? Kant asked himself,—and what really was his answer? *By virtue of a faculty*, which is to say: *enabled by an ability*; unfortunately, though, not in these few words, but rather so laboriously, reverentially, and with such an extravagance of German frills and profundity that people failed to hear the comical *niaiserie allemande* in such an answer. In fact, people were beside themselves with joy over this new faculty, and the jubilation reached a peak when Kant discovered yet another faculty, a moral faculty:—because the Germans were still moral back then, and very remote from *Realpolitik*.—The honeymoon of German philosophy had arrived; all the young theologians of the Tübingen seminary ran off into the bushes—they were all looking for "faculties." And what didn't they find—in the innocent, abundant, still youthful age of the German spirit, when Romanticism, that malicious fairy, whispered, whistled, and sang, when people did not know how to tell the difference between "discovering" and "inventing"!

—FRIEDRICH NIETZSCHE

Modern philosophy aims to explain how mindedness is possible in a material universe. In one of the most provocative statements of this discourse, Wilfrid Sellars claims that internal representations, which are distinctive of mentality, are a theoretical accomplishment, and attributes their invention to a single creative act by "Man himself."[1] This book argues that acceptance of Sellars's conception of mentality as a theoretical accomplishment is plausible only when one rejects his picture of singular creation. Mentality did not arise all at once but developed in distinct phases, some of which can be clearly reconstructed.

Sellars allows that the historical emergence of mentality as a characteristically human mode of self-understanding underwent continuous development in the framework he calls the "manifest image" of man. He conceives of this as a "refinement or sophistication" of the "original" framework, an inductive development of the original image rather than a postulation of supersensory entities explanatory of perceptible things.[2] This book argues that, on the contrary, powers of the psyche are supersensory entities. Consequently, the development of our mode of self-understanding is not continuous but emerges in a punctuated equilibrium. Each stage in the history of mentality consists of a relatively stable picture motivated by certain theoretical considerations. Yet a privileged moment in the overall process of the development of psychology, examined in this book, is that in which psychological appearance emerges for the first time, becoming a theme for reflection and a tool of explanation.

Divisions of the psyche allow for the clustering of various functions under the divided parts. These functions might be explained biologically, as evolved capacities or as complexes of submechanisms. But the grouping of these functions is not primarily a scientific matter. To ascribe mentality to someone is not to attribute to that person a certain evolutionary history or to assume that his body is analogous to my own. But if artificial minds quite unlike my own are possible, similarities in evolutionary history or bodily condition are not essential to mindedness. To ascribe mentality to a person is, rather, to see that person as having an inner life of thoughts and feelings that may only secondarily be communicated.

The explanatory roles of a psychological concept are determined by a specific division of the psyche. Thus, for Plato and Aristotle motivating desires are not associated with freedom and responsibility. For they take desire to be naturally directed toward the good and thus in an important sense constrained rather than free. The connection to moral responsibility develops gradually in the Hellenistic schools and is becoming clearly articulated in Augustine.[3] Freedom, which is initially a political concept of not being under compulsion, is internalized in the psychological concept of will.[4] This new clustering of functions—of choosing between alternatives, acting voluntarily, and so on—under the heading of will gives rise to an understanding human action that becomes sedimented and "obvious." By Descartes's time, possession of free will seems to be a natural way to distinguish human beings from automata.[5]

Psychological inquiry is a special case of scientific investigation. Functional grouping—the clustering of observable activities under a supersensible faculty—cannot be falsified or verified by any single experiment or observation. In Lakatos's terms, the negative heuristic of a psychological theory disallows anomalous phenomena from counting against a grouping. The positive heuristic explains anomalies by making adjustments to the protective belt of assumptions that accompany the psychological division.[6] A research program in psychology renders explanatory resources that give shape to a logic of discovery in which behaviors are taken to be functions of the parts of the psyche according to the favored division.

Theory change occurs when a new division of the psyche becomes progressive, eclipsing a previously dominant but now degenerative program of research. Thus, a psychological research program is falsified when an alternative program offers better explanatory resources at a lower cost. Many philosophers of science think of such costs and benefits in cognitive terms. Endogenous theory change might occur when cognitive costs are reduced, for example, by the application of Occam's razor. In the following pages, I offer an endogenous account of a development in ancient psychology, in which theoretical considerations internal to the Academy drove Aristotle's division of the psyche and resultant grouping of functions. But theory change need not be endogenously determined. Exogenous conditions such as ideological factors, sources of patronage, or intervention by a political regime might also play a role in how we understand the mind.

Are the parts of the psyche discovered or invented? The realist view is that a correct division of the psyche describes things as they are. For the realist, thinking is instrumental to knowledge of objects, so that the psychologist discovers parts of the psyche. In contrast, to use Dewey's terms, the instrumentalist takes thinking to be instrumental to a control on the environment. This control is produced by means of a conceptual division of a "complex situation into assured elements."[7] For the instrumentalist, the psychologist invents the parts of the psyche as controls on what he observes—for example, overt behaviors or reports of introspective experience. Psychological theory is not the mirror of a subsistent nature but an attempt to advance a human aim—namely, to predict and explain behavior. In antiquity, some peripatetic philosophers understood psychological divisions to be defined only in terms of the effects they have rather than divisions in substance.[8]

The question of whether psychological powers are discovered or invented thus depends on whether scientific theories describe reality. But this is typically difficult to decide. As Ernest Nagel observes, a postulated, supersensory object can be said to be real if it is (1) something that would be publicly perceived under suitable conditions, (2) something postulated by a theory that is widely accepted and well supported by empirical evidence, (3) something that enters into an experimental law, (4) something that enters into a (theoretical or experimental) causal law, or (5) something that is invariant under a stipulated set of transformations.[9] For reasons that I hope will become clearer in the course of the argument of this book, the imagination does not obviously satisfy any of these conditions. Attributions of imagination are widely accepted, but because the phenomena described by them are so diverse, it turns out to be difficult to say what imagination is in the first place. Consequently, I speak of the "invention" rather than the "discovery" of the imagination. But it seems to me that to understand the emergence of mentality, we need not a theoretical reconstruction of the science of psychology—if there ever were such a thing—but a historical narrative that puts our own self-conception in relief.

Sellars holds that the postulation of an "inner sequence of events" helps explain human behavior, especially "iffy" facts about what one might do in a hypothetical or counterfactual scenario.[10] The framework built up from this postulation is manifest rather than scientific because internal appearances were not postulated on a neurophysiological basis. Hence the manifest framework gives the methodological conditions for the scientific framework to be developed so that the scientific framework is "methodologically dependent on the world of sophisticated common sense." Yet because it "purports to be a complete image," it emerges as a rival to ordinary psychological description.[11]

But Sellars fails to grasp that the manifest image is itself subject to change through theoretical developments that lie outside of ordinary or "folk" psychological concepts. The same stretch of highway is both the road from folk psychology to the scientific image, and from theoretical science to sophisticated common sense. Postulates can travel in both directions along this two-way street. The picture of rival frameworks is therefore too simplistic to capture changes in psychology. A more apt metaphor would be of a single city with residents following various ways of life in each neighborhood. Neighborhoods are not static but undergo constant changes as residents are born, move, or die. Some areas might be partly rebuilt or expanded, even as the neighborhood retains its identity. And, insofar as they pass into one another, in any given area "there is no last house on the road."[12] Although designating a concept as "folk-psychological" or "scientific" might have a real descriptive or rhetorical purpose in certain contexts, there can be no absolute division between the "ordinary" and "technical" concepts we use to describe psychological phenomena.

This conceptual intertwinement stems from the reflexive nature of psychological descriptions. The concepts of folk psychology are not just part of an inherited theoretical apparatus by which we explain the behaviors of others. They are the terms in which we understand ourselves. The capacity to say what I see, imagine, or think develops apace with my ability to understand other persons as subject of the same attitudes. Thus, general hypotheses such as "every human being has the power of imagination" are insufficient to ground psychological ascriptions. For token thoughts and feelings to be recognizable as such, and thus as candidates for expression, one must take oneself to be the type of thing that can entertain them. A minor premise is required to ground self-ascriptions of psychological terms, such as "I am a finite thinking being" (or, in an etiolated and merely contingent form, "I am a philosopher" or "I am a Hellene"). Real psychological ascriptions are typically enthymemes, in which such minor premises are suppressed and implicit. Nevertheless, they ground every schema of the manifest image, making folk-psychological theory inseparable from its immediate use in telling myself what I am thinking and feeling. Alexander of Aphrodisias was remarkably prescient when he characterized Aristotle's psychology as the attempt to gain self-knowledge.[13]

This book argues that a new neighborhood sprung up with Aristotle's introduction of the power of appearances, which he took to be produced by a new part of the psyche, the imagination. What substantiates my claim that theoretical developments can be absorbed into the manifest framework, and thus into our conception of ourselves? It is that Aristotle conceived of the imagination as the part of the psyche responsible for producing and presenting geometrical objects that are intermediate between sensible particulars and pure universals. Thus, purely theoretical considerations caused him to postulate psychological appearances, which in turn required a new functional grouping of activities. What is needed to understand the development of this new form of self-understanding is an examination of this root concept of internal appearance. This book examines this root concept by showing how imagery was introduced into psychology in fifth-century Athens.

My argument begins with a discussion of the terminology surrounding images and appearances in the Greek literature prior to Aristotle. Phantasia in Plato shares the pejorative connotations of some of Aristotle's descriptions. For both philosophers, this is a power that gives false proportions to what it fabricates. But Plato ascribes this power to things and to art: deception is produced by something external and the psyche is, so to speak, innocent of any epistemic wrongdoing. Aristotle's innovation is to put the power and its products, the phantasmata, in the psyche. The psyche thereby gains the inborn power to make and put on display its own objects. Phantasia thus becomes a psychological power, the imagination.

Methodologically, the book's gambit is to connect Aristotle's psychology, developed in *De Anima* and in the *Parva Naturalia*, to his comments on geometry, which serves as a canonical example of science in his *Analytics* and *Metaphysics*. In connecting these texts, I take his psychology to be a *theory of cognition* meant to accompany Aristotle's *theory of science*. In this view, the psychological works should disclose conditions for thinking the objectivities discussed in his analytics and metaphysics, syllogisms and beings, respectively. However, much geometrical detail is omitted from Aristotle's work, in which he typically assumes that his auditors or readers have explicit proofs and figures in hand or in their memories. Since we are not students in Aristotle's Lyceum, this assumed material must be resupplied from external sources. To achieve this reconstruction, I make use of late antique texts, chiefly Proclus's *Commentary on the First Book of Euclid's Elements*, our main extant source on the history of pre-Euclidean geometry. Otherwise, my interpretations are based mainly on comparing passages within the Aristotelian corpus, and in tracking how Aristotle uses specific words within certain contexts. While it may be impossible to archive mathematical certainty about the true interpretation of any historical text, I hope at least to stimulate dialogue among students of the history of cognition, and perhaps also to persuade some philosophers that the ancients still have much to offer them.

CHAPTER 1

IMAGES *and* APPEARANCES

"Image" may indicate a likeness between two distinct objects. A son is the image of his father insofar as their appearances are similar, even if they are not identical in any single respect. The example of father and son relates to a secondary, reproductive sense in which designating something as an image marks its derivative status. A Roman reproduction of a Greek sculpture, though perhaps indiscernible from the original, is an image in this sense of copy. A third function of the word concerns the role of images as semblances, or feignings of reality. Used in this sense, the image's likeness to the thing it represents is incomplete. Such a self-dissimulating image is not a purely causal reflection of a natural form but must be created intentionally.[1] "Image" can connote one or more of these senses, but its extension is indeterminate. A son is not the image of his father in the way a photograph is. What counts as an image depends on what sort of conversation one is having.

In the earliest Greek sources, an image (εἴδωλον) is associated with the dead. In the *Iliad*, when Achilles has fallen into a deep sleep, his recently deceased companion Patroklos appears to him in a vivid dream. Patroklos warns that Achilles will also die at the hands of the Trojans and implores that their bones be interned together in a single urn. But when Achilles rises to embrace him, Patroklos's psyche disappears in a vapor. Achilles, jumping to his feet and striking his hands, laments that "in the house of Hades, there is a psyche and *eidelon* that have no life in them," and that such an image has hovered over him all night, looking wondrously like (*eikto*) Patroklos.[2] The image here indicates the dead man's similarity, as well as his status as a derivative and mere semblance of Patroklos's living body. As Garland writes of the dead in archaic Greek literature, "Linguistically their insubstantiality is suggested by such words as *eidôlon* (image) and *skia* (shadow), which unambiguously indicate that the dead as perceived by the living were in a very literal sense mere shadows of their former selves."[3] Images are insubstantial but nevertheless objective, insofar as the dream is something *seen* by the dreamer.[4]

The rise of figural representation in art leads to the development of a different concept of the image. Vernant argues that the language of plastic forms is unknown in the Dark Ages, with the verb "to draw" (γράφειν) continuing to indicate as disparate activities as writing letters and painting figures for several centuries. Archaic religious carved idols, the xoana, do not represent deities in their canonical forms but function as religious artifacts, the contemplation of which serves to initiate the viewer into the invisible world of the divinities. The representational form of the idol thus has not yet reached "full autonomy" and "the boundary is still rather fluid between the xoanon and certain symbolic objects that also bestow a particular religious quality on their possessors."[5] Early Greek sculptural productions are not images that depict an external reality but objects of cultic contemplation, which draw the individual viewer into a religiously modified state of consciousness.

The public statue, which arises with the institution of the temple as the house of a god in a city, represents a decisive break from the older tradition of symbolic objects. Divinities are now represented conventionally, for display before a viewing public. Thomas Hobbes connects this to the rise of the state, since "before introduction of Civill Government, the Gods of the Heathen could not be Personated."[6] Vernant agrees with Hobbes, writing: "The statue is representation in a really new sense. Liberated from ritual and placed under the impersonal gaze of the city, the divine symbol is transformed into an image of the god."[7] The publicity of the statues of the gods allows them to function as physical embodiments of state power.

Statues function in a context in which the human body is gaining importance. The representation of human forms in archaic religious art, sometimes misunderstood as "anthropomorphizing" the gods, valorizes the human form as a reflection of the divine, a status that develops apace with the civic and religious significance of athletic competition. The spectacle of struggle among naked, masculine bodies stands in for the relations among cities, serving a civic and religious function, in which victory consecrates the victor.[8] Nicholson argues that athletic victory serves as a sign both of virtue and of divine favor.[9] From the mid-sixth century to the mid-fifth century, victory monuments justify aristocratic political power to a broader public. Indeed, statues of athletic victors erected in temples and other civic spaces were sometimes thought to have special powers, such as the ability to cure fevers.[10] The valorization of human form thus corresponds to a political culture in which aristocracy is represented as a divinely sanctioned form of rule.

The mimetic function of the image develops initially in funerary sculptures, which in the late sixth century were erected to reproduce the lost beauty of the dead. In the late archaic period, this practice loses its religious meaning but gains a new significance, in which statues serve as imitations or copies of a living body. Vernant writes, "For the image to acquire the psychological significance of a copy that imitates a model and gives the spectator an illusion of reality, the human figure must have ceased to incarnate religious values; in its appearance, it must have become in and for itself the model to be reproduced."[11] This loss of ritual significance by symbolic objects and of religious meaning by the human body gives rise to the mimetic function of statuary, in which the image is a mere semblance of a freestanding reality.

The mimetic function of artistic images gives rise to an anxiety, portrayed dramatically by Euripides in Admetus's speech to his dying wife, Alcestis. "Represented (εἰκασθέν) by the hand of skilled craftsmen, a figure shall be laid out in my bed. I shall fall into its arms, and as I embrace it and call your name, I shall fancy (δόξω), though I have her not, that I hold my dear wife in my arms, a cold pleasure, to be sure, but thus I shall lighten my soul's heaviness. And perhaps you will cheer me by visiting me in dreams. For even in sleep it is pleasant to see loved ones for however long we are permitted."[12] Here the funerary image is a replica of Alcestis, built to replace her in the bed of her bereaved husband. The state of fancy that image stirs in Admetus is not unrelated to dreaming of his deceased wife: both are fleeting and unreal but somehow relieve his soul of the burden of her death. Gombrich characterizes this image as inculcating a "dream for those who are awake." For Gombrich, this discovery of a "twilight realm" of waking dreams marks the beginning of an emancipation of consciously constructed fiction from myth.[13] This eventually leads to the Platonic anxiety about the untruthfulness of images.

These considerations about the development of the concept of the image may be supplemented by considering some social functions possessed by images in the sixth and fifth centuries in Greece. In what Steiner calls a logic of replacement, an image depicting an object could come to stand in for the object itself. For instance, the manufacture of a statue of a dead person might appease his disquiet spirit by allowing for its symbolic reinternment in a body. In this case, the statue stands in for the absent body. Thus Steiner reads Herodotus's story concerning the deceased Aristeas, whose body disappears but who returns as an apparition (φάσμα) to demand that a monument be erected in Apollo's grove, to assume a power of statues "to actualize the individuals' power at the site and to make it continuously accessible to those who have erected the monument."[14] The continuance of this power is made possible through the replacement of the living body of Aristeas with his memorial statue. A related, political function of images is suggested by an inscription recording an explicit social contract of seventh-century Greek colonists in Libya. According to this inscription, the colonists burned wax images (κολοσσὸι) while stating, "May he who does not abide by this agreement, but transgresses it, melt away and dissolve like the images, himself, his seed, and his property." As Steiner points out, "the efficacy of the ritual depends on the statuettes' capacity to re-present the oath takers, and not merely to symbolize but rather to prefigure the perjurer's eventual fate."[15] Finally, there is a fixative function of imagery, in which an image (εἰκών) traps a restless ghost or repels it from the site it haunts, such as the place in which the person had died. Crucially, such an image is made not of the original living body but of the ghost itself, exorcising the invisible apparition by giving it substance.[16]

These athletic, religious, and spiritualistic functions of produced images did not escape criticism by philosophers. Empedocles describes artists' paints as narcotics (φάρμακα) by which they supply a whole world of likenesses.[17] One finds a more iconoclastic sentiment, directed against both bodily valor and the religious significance of the human figure, in the late archaic fragments of Xenophanes of Colophon. The sixth-century wandering poet complains that having good boxers, pentathletes, wrestlers, fast runners, "not for that reason would the city be any better governed."[18] By pointing out that athletic accomplishment does not lead to better governance, Xenophanes criticizes the games and suggests that athletic *agon* is not a suitable model of civic virtue. Xenophanes also writes against the anthropomorphized conception of the gods: "There is one god, greatest among gods and human beings, not at all like mortals in form, nor yet in mind."[19] The greatness of god, far from stemming from his likeness to humankind, is understood in his dissimilarity to us. Xenophanes worries that "mortals believe that gods are begotten and have clothing, voice, and body like their own."[20] Such characterizations are ridiculous since "Ethiopians say that their gods are snub-nosed and black; Thracians say that theirs have blue eyes and red hair."[21] Xenophanes's primary reason for rejecting this anthropomorphic religious imagery is its arbitrariness, which degrades the divine. In portraying the gods as snub-nosed and dark-skinned, or with red hair and blue eyes, artistic representation misses the crucial point that divinity consists in what is superior to and not shared by mortal men.

Xenophanes's polemic against the anthropocentrism of representational art, at its most extreme, becomes farcical. Supposing that the artistic representations of divinity are based on the bodies of those who make them, he considers a situation in which nonhuman animals could make their own representations of the gods.

> If oxen and horses and lions had hands,
> And could draw (γράψαι) and bring about works (ἔργα τελεῖν) like men,
> Horses like horses, and oxen like oxen,
> Drawing forms of the gods (θεῶν ἰδέας) and giving them bodies (σώματ' ἐποίουν),
> These would be just like the one that each of them has.[22]

If horse- and ox-shaped gods are ridiculous, then portraying gods in human form is equally absurd. This critique of religious imagery is directed at the image's function as a reproduction of the familiar, in this case the bodies of the imagined animal artists. Just as the animals, when given hands, bring to completion drawings that reflect their own bodies, the work of human hands represents not the invisible, true god but the visible human form. Allegedly sacred images are revealed as cheap copies of human bodies, mere semblances of the divine.

Dressed in attractive poetic language, Xenophanes's anti-anthropomorphizing arguments are severe and comprehensive. Not only does he attack the civic function of the body and the religious meaning of statuary, but poetry becomes an object of critique. Xenophanes testifies that "both Homer and Hesiod ascribed to the gods all things that evoke reproach and blame among human beings: theft, adultery, and mutual deception."[23] The greatest poets put the gods in human attitudes, attributing to them not just human shapes but also human flaws. Xenophanes reveals the absurdity of this error. Thus the relation between truth and graphical representation, whether in artistic drawing or poetic writing, is a point of contention a century and a half before Plato, who refers in his own work to the "ancient quarrel between philosophy and poetry."

Plato follows Xenophanes in criticizing Homer, the first of the "initiative poets" (*Pol.* 607b, 600e), and in focusing his polemic on the untruthfulness of imitative poetry and its consequent ability to replace beings with mere semblances. Yet Plato's polemic in the *Republic* radically transforms Xenophanes's iconoclasm. Plato's opposition to oral culture required him to distance himself from the language of the Homeric tradition.[24] Thus, Havelock argues, the movement of the psyche in Plato's work from the many to the one or from "becomingness" to "beingness" corresponds to "a conversion from the image world of Epic to the abstract world of scientific description, and from the vocabulary and syntax of narrativised events in time towards the syntax and vocabulary of equations and laws and formulas and topics which are outside time."[25] If for Xenophanes the religious image was problematic in that it corresponded to the prejudices and false conceptions of its producers, for Plato the image has a peculiar metaphysical standing as an imitation, a status defined in relation to the unchanging reality it purports to represent. Tying this disparagement of the image to Plato's elevation of the idea, Havelock concludes that "it is fair to say that Platonism at bottom is an appeal to substitute a conceptual discourse for an imagistic one."[26] If this is correct, then Plato's understanding of figural representation constitutes a complete reversal from that of the early archaic period: whereas the image once served the ritual function of drawing its perceiver into the higher world of the gods, for Plato the image becomes a mere semblance of a semblance, at "a third remove" from nature (*Pol.* 597e cf. 602c). Plato's rejection of the image as faithful representation of reality leads him to suggest that beings must lie outside of the perceptible realm altogether. Plato's rejection of images is therefore bound up intimately to a conception of appearances as misleading products of perceptual intercourse with the world.

How is Plato's critique of imitative art related to his conception of appearances? In *Republic* X, Socrates explains to Glaucon how images derive from appearances.

> "Consider (σκόπει) this: about what is painting (ἡ γραφική) made (πεποίηται) in each case? Does it imitate (μιμήσασθαι) being as it is (τὸ ὂν ὡς ἔχει, μιμήσασθαι) or the apparent as it appears (τὸ φαινόμενον ὡς φαίνεται)? Is it an imitation of appearance or of what truly is (φαντάσματος ἢ ἀληθείας οὖσα μίμησις)?"
>
> "Of appearance," he said.
>
> "Then imitative art (ἡ μιμητική) is far from truth, it seems, and because of this can enact everything, so that it hangs onto an aspect of each thing, and this is the image (εἴδωλον). For example, we say that the fig-

> ure painter will paint (ζωγραφήσει) from life a cobbler, a carpenter, and other craftsmen." (*Pol.* 598b)

There are two distinct concepts at work here, which are sometimes misleadingly conflated or treated as interchangeable in translations. The image is a human "production" or "imitation" of a natural form. The appearance is the product of a perceptual episode. Plato explains that images derive from appearances. Appearances, however, arise not through art but by our nature as embodied creatures that inhabit certain perspectives and perceive particular features of objects such as magnitudes, shapes, and colors.

The argument of the passage begins with the premise that what is imitated in figurative art is not an idea but a sensible particular, a "thing." Plato elsewhere considers as examples the sun, things in the sky, things on the ground, people, animals, manufactured items, and plants (*Pol.* 596d–e). The question is whether the painter imitates the thing as it truly is, or an appearance of the thing. The interlocutors agree that imitative art copies only a thing's appearance, providing the basis for the view that images are "further" from a thing's nature than the perceptual episode in which the thing appears. Intermediate between the artist's image and the reality it purports to represent is the appearance, which is imitated by the artist. The comparative freedom of imitative art, its ability to produce everything, hanging on to only an aspect of what is real, therefore corresponds to its distance from being. Like the mirror that merely reproduces a thing's appearance, the figure painter produces an appearance of an appearance (*Pol.* 596d–e). Plato concludes that "the maker of the image knows nothing of the being, but only of the appearance" (*Pol.* 601b). Rather than seeing nature as it is, the painter inhabits a single and partial perspective, which is merely reproduced in the painting.

Plato's argument is sometimes criticized for assuming that art must take all its suggestions and materials from nature.[27] But whatever one thinks of that premise, it is clear this passage takes images to derive from appearances. The language of appearance used here has different parts: *phainomenon* indicates an object's appearance in a perceptual or quasi-perceptual episode; *phantasma* indicates the particular episode in which this object is given. Plato thus speaks of something's being given "as an appearance" (ὡς φαίνεται) to indicate the unreality or unreliability of the way in which it is presented. Plato argues that there must be a nonrational power of the psyche set over appearances (*Pol.* 603a). For Plato, this power must be perceptual, since false appearances arise paradigmatically from episodes of visual error. A magnitude's appearing unequal from nearby and from afar, something convex appearing concave due to the colors and shading of paints, a straight stick appearing bent in water, irregularities in size, shape, and color—are all cases of such misperception. Plato uses only visual examples but insists that appearances exist "not only in the case of visual objects, but in objects of hearing too" (*Pol.* 601b). Thus, Plato takes appearance to be the product of a perceptual episode that causes a confusion (ταραχή) in the psyche (*Pol.* 602c–d).

Turning to the arts, Plato is concerned to show how images exploit this natural confusion, and hence affect our nature "as if by witchcraft" (*Pol.* 602d). The central feature of Plato's critique is directed to the "images of virtue" (*Pol.* 600e) in tragic poetry, which held so much political and religious importance in his time. Like appearances, images for Plato are not internal states of the psyche but are external, having the *effect* of throwing the psyche into confusion. For Plato, while images of virtue have psychological relevance, they are not generated in the psyche. Rather, irregularities of size, shape, and color themselves are taken to be objects of perception. Consequently, for Plato, visible things include both "the kinds of things that are about us," such as animals, plants, and manufactured goods, and "semblances" (εἰκόνες), which include shadows, appearances (φαντάσματα) in water, reflections (φανὰ) on hard, smooth surfaces, and "all the things like that" (*Pol.* 509e–510a). Whatever the metaphysical status of visible objects in the *Republic* may be, Plato takes it that appearances—*phantasmata*—are perceptual errors or irregularities that mislead and confuse the psyche, which are not themselves states of the psyche. In conceiving appearances and the images derived from them in this way, Plato suggests that when one misrepresents how things are, the deception is to be attributed not to one's psyche but to the world that presents itself to one.

I have been arguing that phantasmata in the *Republic* are not psychological states but products of observing the changing world. Other texts in the Platonic corpus corroborate this nonpsychological interpretation. Consider the section of Plato's *Sophist* (235d–236c) in which the Athenian Stranger divides imitative art into *eikastikē technē*, which reproduces the true proportions of an original paradigm, and *phantastikē technē*, which distorts those proportions to present things to the viewer in a certain way. These terms evidently have a different sense here than their cognates have in the *Republic*: εἰκαστική means "copying" while φανταστική means "perspectival." The former techniques are demanded in constructing a scale model of a building, where accuracy is crucial, whereas the latter is used to great effect in Michelangelo's *David*, in which the unnaturally large head *looks* proportional and thus beautiful to those who view it from a couple of meters below. Here, too, however,

Plato employs *phantastikē* as a descriptor, not of the psyche of the viewer but of the artist's technique and resultant work. Michelangelo's *David* is *phantastikē* in that it is apt to deceive one in a pleasing way. This deception originates in Michelangelo's technique, not in the viewer's psyche.

There is one passage in Plato's *Philebus* (38c–39c) that seems to propose a psychological theory of mental imagery. When one sees something dimly in the distance, one might wish to discern (βούλεσθαι κρίνειν) what it is (38c). For example, one might observe a figure (φανταζόμενον) under a tree. One then performs the identification of the figure by asking a question of oneself ("What is it?") and by answering that question ("It is a man."). When the question is answered correctly, one forms a correct opinion about what one perceives. This process of identification through the asking and answering of questions suggests that the psyche is like a book, in which verbal descriptions are recorded. Plato understands these verbal descriptions to be affections that appear in the perceiver (παθήματα φαίνονταί) due to the work of perception and memory. The conclusion of this line of thought is the striking metaphor that whenever one forms a true or false belief, there is a scribe who writes words in the psyche (παρ' ἡμῖν γραμματεὺς γράψῃ) (39b). The description of mental imagery follows as an addendum to this account of perceptual judgment: when the words are written, another craftsman (ἕτερον δημιουργὸν) in the psyche produces an illustration of the scribe's description. This other craftsman is a scene painter (ζωγράφον), who draws (γράφει) images (εἰκόνας) in the psyche after the scribe records the words. Both the verbal descriptions and the images that accompany them represent past and present objects, as well as future possibilities and expectations, such as hopes for future pleasures (39d-e).

This passage is sometimes considered to employ a concept of imagination. But the psychological images invoked in this passage are taken to emerge *after* the perceiver has apprehended an appearance and given a verbal description of that appearance to himself. So, the images of Plato's account cannot play a causal role in cognition but merely reflect the truth or falsity of the antecedent phantasmata. Moreover, the phantasmata are understood as presentations from the world, which impress themselves on the perceiver, rather than products of a psychological power. Even if one lacked a painter in one's psyche, the world might still present a thing to one falsely, leading to a false verbal description. Moreover, Plato treats appearances and images as distinctive phenomena, which cannot be equated. Thus, it seems that Plato does not here postulate a power of the psyche that produces of appearances but consistently ascribes appearances to the world. Nevertheless, Aristotle's psychological theory is evidently informed by Plato's writings and by ideas he encountered during his twenty years in the Academy. In this sense, one can say that Aristotle's theory of imagination as a psychological power stems from taking Plato's craftsman to have mastered the *phantastikē technē*.

This point can perhaps be strengthened by comparing it to another pre-Aristotelian psychological doctrine that might be taken to offer an incipient account of the imagination. According to later reports, the atomists have a materialist theory of sense perception, in which images play a central role. On this theory, both thinking and visual perception consist in the apprehension of images (εἴδολα). Democritus employs the technical term *emphasis* to demarcate this sort of image, as opposed to, for example, the images produced by painters. According to this theory, images are able to represent objects to a subject because they share a form with visible objects. However, upon close inspection, it turns out that these images are not psychological items. Rather, they are impressions in the air, which arise between the object seen and the sense organ.[28] These images continually stream off of visible objects, and are hence from "outside" the perceiver.[29] Thus, while there is a tradition preceding Aristotle that understands perceptions to be representational or depictive, the agency responsible for producing and presenting them is invariably something outside the psyche.

Why do ignorant or inexperienced people tend to be deceived more often and more easily than those who are experienced and knowledgeable? The answer, according to this nonpsychological interpretation of Plato's account of images and appearances, is that everyone is subject to false appearances. The straight stick in water looks bent and Michelangelo's *David* seems proportional even to experienced and knowledgeable people. The difference is that the experienced or knowledgeable person, unlike his foil, has learned not to trust all appearances, and thus doubts that the world is always as it appears to be. Plato's discussions of images and appearances aim to dispel the misconception that things are always as they seem. In these dialogues, Plato offers a vision of philosophy as the path of liberation from the tyranny of seeming, an escape into reality.

If representational art and tragic poetry elicited a reactionary response from Plato, the Sophists provided a more direct challenge to the possibility of knowledge of an independent reality. Consider Protagoras's claim in his book *On Truth* that "of all things, the measure is man, of the things that are, that they are, and of things that are not, that they are not."[30] This maxim is humanistic in the sense that it makes each human being the criterion of what is and is not. It is relativistic in that it sets down appearance alone as what makes the truth for each person.[31] Protagoras thus proposes to replace the concept of truth with a concept of utility, according to which knowledge and expertise are to be judged by their usefulness to mankind.[32] Thus, Protagoras could describe his doctrine as "the art of politics" since by his criterion of utility he could teach political expertise.[33] Sophistical humanism and relativism consequently dispenses with any absolute or superhuman criterion, not only for truth but also for beauty and virtue, and attempts to reconstruct them on non-theological, purely human grounds.[34] Plato's critique of Protagoras is a response to a crisis in Greek thought brought on by the rise of sophistical rhetoric.[35]

Plato resists this humanistic and relativistic doctrine by arguing that something's appearing to be the case is insufficient for it actually being the case. In the *Theaetetus* (152a–c), he interprets Protagoras's humanist maxim to be equivalent to the view that any given thing "is to me such as it appears to me and is to you such as it appears to you." This entails that the perception that something is hot or cold is always veridical, so that knowledge is appearance (φαντασία) and perception. Protagoras must consider the question of what something is like, apart from what it is like for one person or another, to be senseless.[36] However, according to Plato, Protagoras undermines his own argument. If each person has the sole authority to decide what he ought to think, there is no reason to give credence to any argument, even to arguments of experts such as Protagoras himself.[37]

Regardless of whether this refutation is valid, it rests on a rejection of the view that appearance and perception are sufficient for knowledge. Knowledge, it seems, requires something more than perceptions and true opinion. Indeed, Plato's epistemology can be understood as attempting to establish, against the Sophists, a disjunction between reality and appearance. Reality is grasped by the sciences. But how something appears to one, how it seems, how it looks, is insufficient for knowledge. An unbridgeable chasm appears between the Sophistical enterprise of producing conviction in an audience and the philosophical and scientific project of gaining true, scientific knowledge of reality. Consequently, while the move toward formalization in mathematics antedates Plato's philosophical reflections, scientific geometry emerges in Plato's time as a privileged field for a philosophical battle between sophistry and science.[38]

The Sophists themselves lead the attack against the geometers. Protagoras claims that when a straight line intersects a circle, they touch not at a single point but across a segment (*Met.* 997b35–998a4). Thus, a geometrical proof that takes intersection at a point as a premise must be false. Indeed, in any drawn figure, a straight line always appears to intersect a ring along an extended segment. If all appearances are true, then it follows that the line in fact intersects along a segment, refuting the geometer. Another sophistical refutation of the geometers is reported by Aristotle (AnPr. 49b35–37, AnPo. 76b39–77a3, *Met.* 1078a17–24, 1089a21–25). Suppose a geometer draws a line, asserts "this is a foot," and demonstrates a geometrical proposition. The objection is that since the unit line is not a foot long, the proposition rests on a false premise. The proof is refuted on the nongeometrical grounds of what is perceived by the objector. Yet another example, due to Aristippus, criticizes mathematicians for not explaining the cause of their results (*Met.* 996a29–31). Because geometers cannot say what agent makes a proposition true, they do not possess real knowledge.

These objections amount to a challenge to philosophy to establish grounds, outside of perceptual appearance, for the capacity of geometry to give knowledge to its practitioners. Yet a return to the early view of geometry as a practical art of earth measurement is out of the question. Consider Herodotus's report that geometry was invented in Egypt as a way to calculate taxes. When the Nile encroached into previously surveyed parcels of land, a method was needed to account for the extent of the loss, and thus to calculate the new, proportionally reduced tax on the holding.[39] This is of no avail against the Sophists, whose objections are directed against the epistemic status of geometry. For the calculation of the taxable area would come down to how it appeared to the surveyor who made the estimate, rather than to any fundamental scientific principle. What is needed was not the return to an understanding of geometry as a craft of calculation but an independent grounding for a truth of geometry, a scientific ontology that deflects objections made on the basis of a perceptual appearance.

Plato's response to this problem is to postulate an independent foundation for geometrical knowledge, which divorces geometrical results from any grounding in appearance. Although he may draw a figure, for example, of a line intersecting a circle, the geometer directs his attention not to the perceptible line and circle but to the "line itself" and the "circle itself." The latter, indivisible figures, cannot be seen "except by means of discursive thought" (*Pol.* 511a1). Consequently, Plato argues in the *Republic*, insofar as geometry is knowledge about what always is, and not of what comes to be and passes away, education in geometry is beneficial, turning the student's soul toward truth and being (*Pol.* 526e, 527b).

Plato gives a powerful example of the difference between reasoning about perceptible figures and reasoning about intelligible figures in the *Meno* (75b–76a), in a passage in which Meno asks Socrates to define shape. The first, "naive" definition proffered by Socrates is that shape is the only thing that always accompanies color. The second, "technical" definition, prompted by Meno's objection that the first assumes one must already know what color is, is that shape is the limit of a solid. The latter makes use of a general concept of limit and explicitly draws on the geometrical concepts of surface and solid, of which Meno has sufficient knowledge to understand the definition. The difference between the naive and technical definitions is that the former assumes acquaintance with the particular objects of perception while the latter proceeds entirely from general concepts of geometry, and requires no particular content.[40]

The difference between these two definitions is mirrored in Plato's denial of the possibility of knowledge of sensible things (*Pol.* 529b). The philosopher is concerned with objects of the race of the intelligible and who is able to grasp what is "always the same in all respects," while the nonphilosopher "wanders among what is many and varies in all ways" (*Pol.* 484b4–7). This corresponds to a division between "theoretical" arithmetic that counts pure units and "practical" arithmetic that counts heterogeneous units.[41] For example, commercial transactions require practical calculation (*Pol.* 525c-d), which is concerned with heterogeneous objects such as cattle or soldiers. Its theoretical counterpart, pure arithmetic, concerns only things that are homogenous or perfectly like one another (*Philebus* 56d–e, cf. *Met.* 1080a20–23).

According to Aristotle, Plato was made aware of the flux of perceptibles from his acquaintance in his youth with the Heraclitan teachings, which, when combined with the dialectical method of Socrates, led him to conclude that it is impossible to give common definitions of perceptible things, since they are always changing (*Met.* 987b6–7). Since it makes no reference to perceptible things, the geometrical definition of shape can serve as the hypothesis of reasoning that gives knowledge of what always is, while the naive definition does not give knowledge but concerns heterogeneous things, whose colors are Empedoclean effluences "commensurable with sight and perceptible by it" (*Men.* 76d). In other words, the definition of perceptible shape speaks only about the race of the visible while the definition of geometrical shape concerns the race of the intelligible (*Pol.* 509d8).

This distinction between changing sensible particulars and permanent intelligibles gives rise to the ontological question concerning mathematical objects—namely, whether they are to be identified with forms, or whether they are neither forms nor sensible particulars, but are "intermediate" between them.

The doctrine that mathematical objects inhabit a third domain, between perceptible particulars and unchanging forms, is a matter of some debate. Tait argues that there is "no room for intermediates" in Plato's *Republic*, where there is a strict division between the opinable (δοξαστόν) and the knowable (γνωστόν). Plato takes powers of psyche that accomplish the same thing to be the same power, while if they accomplish different things, they are different powers (*Pol.* 477c–d). It follows that knowledge and opinion are different powers, and thus that the knowable is not the opinable (*Pol.* 478a–b). Since the forms are the objects of knowledge, it is impossible that there be intermediate entities besides them, which are also objects of knowledge.[42]

Yet insofar as Plato makes a distinction in the upper part of the divided line between discursive thinking and intellect, he opens the possibility of reasoning about knowable objects in two distinct ways, either from hypotheses to conclusions or from hypotheses to a nonhypothetical first principle (*Pol.* 510b). Since we form propositions and make inferences in virtue of discursive thinking rather than intellect, geometry must have a hypothetical method. Thus, Cornford argues that the concept of hypothesis arose originally in a didactic situation, in which a teacher's assumptions were "put to the learner in the process of instruction." By accepting such assumptions, the learner comes to appreciate how the teacher's argument hangs together, and in this sense hypothesis rests on a "kind of agreement between the teacher and pupil."[43] Before the Euclidean codification of geometry, which reduces the assumptions necessary for geometry to the fewest possible, this would have been the natural way for geometers to proceed.

Following Aristotle (An.Po. 70b23–34) in distinguishing hypotheses relative to the learner from those that serve a technical function as principles of a science, Cornford holds that hypothesis typically has the technical ("absolute") sense in passages in the *Republic* discussing the sciences (510c) but also retains its original sense in certain instances (533c).[44] Likewise, Lee identifies technical hypotheses with Euclidean postulates, as suggested by Proclus.[45] In any case, the divided line represents intellect as nonhypothetical, so that what is grasped is known in a more fundamental sense than what is accepted merely hypothetically.

Even if the relation between the species of cognition represented in the divided line remains mysterious, it is clear that the division between the knowable and the opinable in the *Republic* does not foreclose the possibility of different types of reasoning about the objects of each realm. Reeve argues that the intelligible contents of discursive thinking cannot be forms, both because the powers set over each section of the line differ insofar as they are set over different things, and because the powers are said to differ in clarity.[46] Plato identifies the geometer's state of mind with discursive thinking, which is intermediate between opinion and intellect. Reeve calls these intermediate objects "figures."[47]

How these figures relate to the sensible and the intelligible parts of the divided line remains unclear. The line could be taken to divide either objects, methods of procedure, deductive or intuitive movements of thought, or states of mind.[48] For example, Hackforth argues that the purpose of the divided line is not to contrast objects but methods of attaining knowledge about objects.[49] Thus the division of the intelligibles does not postulate mathematicals as intermediate but sets in opposition the imperfect method of hypothesis with the perfect method of dialectic. Hackforth's interpretation hinges on taking the "ascent" from hypotheses (*Pol.* 533c–d) as not indicating the shift of the inquirer's interest from one set of objects to another but rather in a perfection of the method in inquiry.[50] There is no obvious way of dividing the ontological, epistemological, and logical interpretations of the line according to criteria given in the *Republic*. Hence while Plato's procedure is surely "methodological," in that it opposes the activity of the mathematician to that of the dialectician, Burnyeat is right to observe that the *Republic* leaves the question concerning the ontological status of mathematical objects "tantalizingly open."[51]

Cross and Woozley find the purported doctrine of intermediates to promise a remedy for an important difficulty.[52] Suppose that two triangles on the same base between the same parallels are proved to have an equal area. The triangles cannot be the sensible triangles drawn by the geometer, since his thought is not directed toward these images but are mere aids in making his demonstration (*Pol.* 510d-e). But nor, if forms are unitary, can the triangles be the form of the triangle, since the creator made each thing one in nature (*Pol.* 577d). This dilemma would seem to be resolved with the postulation of intermediate objects, since the figures grasped by discursive thinking would be extended and multiple like perceptibles, allowing for reasoning about multiple triangles, but would also be unchanging and perfect, so that their areas could be exactly and hence truly equivalent.[53] Yet there is no clear reference to intermediates in the *Republic*, nor in any other dialogue. Considering just the textual evidence of the Platonic corpus, they conclude that the objects of the third section of the divided line must also be forms.[54]

The concern of these twentieth-century scholars to refrain from attributing anything to Plato not explicitly contained in the text of the dialogues was not shared in ancient times. Aristotle's attribution to Plato of the view that mathematicals are neither sensible things nor forms but are "in between, differing from the sensible ones in being eternal and unmoving, and from the forms in that there are many of them alike; but each form is itself singular" is the source of the doctrine of intermediates (*Met.* 987b14–18, cf. 992b15–18, 1077a9–14). Aristotle makes use of this doctrine in his own division of theoretical philosophy—into natural, mathematical, and theological branches—which assumes an intermediate status of mathematical beings between the physical and the divine (*Met.* 1026a19). Similarly, Aristotle writes that "Plato posited the forms and the objects of mathematics as substances, and as a third the substance of sensible bodies" (*Met.* 1028b19–21).

Proclus also characterizes mathematical statements as unchanging, stable, and incontrovertible but as concerning entities that are neither simple, incomposite, nor indivisible (*InEucl.* 3). Thus mathematical procedure is discursive and treats mathematicals as extended, giving different principles for different objects. Such propositionality entails that mathematical judgments are not made by intellect and that mathematicals are *dianoeta* ("understandibles") (*InEucl.* 10). Proclus holds that the threefold division of being into intelligibles, mathematical things, and physical things is Pythagorean in origin.[55] For Aristotle and Proclus, at any rate, Plato's splitting the race of the intelligible into discursive thinking and intellect in the divided line posits two kinds of intelligibles, both of which are independent of perceptible things.

Although the doctrine of intermediate mathematicals is not explicitly endorsed in any of Plato's written works, it sheds light on the positive methodological project of *Republic* VII. Plato there makes a distinction between the pure mathematical sciences of arithmetic and geometry, and the sciences of motion whose objects are grasped by perception since, "as the eyes fasten on astronomical motions, so the ears fasten on harmonic ones" (*Pol.* 530d). Both of the latter sciences are flawed as they are currently practiced since they take the observed motions to be real (*Pol.* 528d–e). Yet observed motions "fall short" of the "true" motions of astronomy, the latter of which "trace out" geometrical figures. Thus, Plato advocates a change in the methodology of astronomy and harmonics from a broadly observational one to one based on geometry, in which physical motions are treated as instantiations of ideal, unmoving geometrical structures (*Pol.* 530b–c, 531b–c).[56] Underlying what comes to be and passes away is what invariably exists. The sciences of things in motion, if they are to render knowledge at all, must render it in geometrical form, since geometry is "knowledge of what always is" (*Pol.* 527b). While astronomy concerns things that are in motion, it should describe these objects with geometrical figures, rooting them in stable and unchanging truths. By rendering statements about the world of becoming in the language of being, the validity of inferences is guaranteed. Plato proposes, in other words, that the natural scientist should approach each phenomenon as an instantiation of intermediates and should direct his attention to the exact properties of those mathematical structures in making further inferences.

While Plato did not explicitly state the doctrine that mathematical objects are intelligible beings intermediate between the forms and perceptible things in his writings, this doctrine was up for discussion in his Academy. According to Dillon, Speusippus derived the universe from two principles. The One, which Speusippus compared to a seed, is the principle that is "before" being. Multiplicity is a fluid diversity that facilitates division into beings and allows the One to be expressed.[57] The result of this division is a "multi-layered universe" that Aristotle satirized as a "stringing out" of being.[58] Among the beings were layers of mathematicals that were taken to be intermediate between the principles and the animate and inanimate bodies called the "fourths and fifths" (54–55). If Dillon's reconstruction is correct, Speusippus may be considered to have multiplied the intermediates. Indeed, Aristotle attributes to Speusippus further intermediate ontological categories (*Met.* 1028b21–24).

Aside from constituting a topic of debate within the Academy, the doctrine of intermediacy puts into perspective how Plato thought that sciences of motion could be made intelligible through geometrization. An observed phenomenon could be reduced to a proposition of geometry, and then used as the hypothesis of further geometrical demonstrations. Because Plato conceives of geometry as a science of intelligibles that are not forms, he characterizes geometry as known by discursive thinking. Socrates says to Glaucon that "you seem to me to call the state of the geometers discursive thought but not intellect, discursive thought being intermediate between opinion and intellect," to which Glaucon replies, "your exposition is most adequate" (*Pol.* 511d). Although the adequacy of this assertion is evidently not as obvious to modern readers of Plato as it was to Glaucon, it is clear at least in the *Republic* that geometry is supposed to operate by taking hypotheses as principles and making deductions from them. Only dialectic, which ascends from hypotheses to the indubitable knowledge of intellect gives genuine knowledge of first principles. Aristotle similarly characterizes geometry as a science that depends on but does not give hypotheses (*Met.* 1005a10; cf. *Meno.* 86e–87b, *Pol.* 510c–d). Aristotle's assertions about the intermediacy of geometrical objects thus fits well with Plato's division between the sciences that study motionless things, arithmetic, and geometry, and those that study things in motion, astronomy, and harmonics.

This ontological distinction between perceptible objects and mathematicals grounds Plato's methodological program. By focusing on the geometrical and arithmetic properties of astronomic and harmonic relations rather than their sensory properties, Plato suggests natural sciences can be raised from the perceptible realm to the intelligible. But if the doctrine of intermediates allows for the banishment of motion from the sciences, a philosophical quandary arises. Given that neither perception nor the other powers can grasp the intelligible intermediates, how are they grasped? The next chapter argues that this question motivates Aristotle's invention of an altogether new power of the psyche—the imagination.

CHAPTER 2

POWERS *of the* PSYCHE

A puzzle presented itself to Aristotle and the members of the Academy in his generation. While perception grasps only material particulars, I show how geometrical knowledge depends on the ability to "produce" and "see" a multitude of immaterial objects. It was necessary to make a division of the psyche that can explain which power grasps these objects.

Plato and Aristotle both deny the identification of perception, which grasps changing material particulars, with thinking, which grasps unchanging universals or ideas. A geometrical demonstration is just a series of statements that follow logically from some premises, whether these are assumed hypothetically or ultimately justified by intellect. One might suppose, then, that geometrical knowledge falls on the side of thinking of unchanging universals.

A difficulty arises from the consideration of the practice of geometers, in which geometrical objects are treated as if they were subject to change. Actual geometrical proofs, before the strictly discursive demonstration given in sentences ordered to deduce the proposition, involve the construction of a figure from which the deduction is subsequently made. Plato already noticed this in the *Republic*, when Socrates explains to Glaucon that the constructive process of alteration in geometry is not really a part of the geometer's procedure but nevertheless necessarily enters into a coherent account of geometry.

> "Now no one with even a little experience (ἔμπειροι) of geometry will dispute that this science (ἡ ἐπιστήμη) is entirely the opposite of what is said about it in the accounts of its practitioners."
>
> "How do you mean?"
>
> "They speak about it absurdly (γελοίως), though they cannot help it (ἀναγκαίως), for they speak as if they were practitioners (πράττοντές), and all their accounts refer to doing things. They talk of "squaring" (τετραγωνίζειν), "applying" (παρατείνειν), "adding" (προστιθέναι), and the like, whereas the entire subject is pursued for the sake of knowledge." (*Pol.* 527a1–b1)

Though geometry is supposed to attain knowledge of immutable, eternal, and unchanging objects, its method is apparently productive, constructing objects from other, more primitive objects and making deductions from those constructions. The paradox at the heart of the academic project of making geometry scientific is thus how to maintain an ontology of ideal, unchanging, geometrical objects and, at the same time, perform operations on those objects that "alter" them.

Plato's exposition of the tension between the intelligible structure of the permanent geometrical objects and the apparently productive activity of the geometer rests on three observations. First, as a science, geometry must reason only about unchanging objects, discursively treating but making no alterations to these intelligible entities. Second, geometers' own reasoned explanations of their procedure necessarily involve the language of production, insofar as they use verbs to describe activities of construction such as squaring, applying, and adding. Finally, the resulting paradox between permanent, unchanging objects and the impermanent, changing figures produced and altered by the geometer is ineliminable—talk of alteration is necessitated by the nature of geometrical proof. As Cherniss puts it, the "contradiction between the language and purpose of geometry" is unavoidable.[1] The paradox of geometry is that geometrical reasoning, insofar as it must be separated from perception, always concerns permanent figures intermediate between changing perceptibles and unextended forms, but geometrical operations also invariably involve producing apparent alterations among these permanent figures. Geometrical objects are supposed to be immutable beings, yet they are treated as if they are merely becoming, coming to be, and passing away at the whim of the mathematician.

While the ineliminablity of productive vocabulary in descriptions of geometrical procedure is already a philosophical puzzle in Plato's writings, in the Academy it becomes a topic of debate in the theory of science. Proclus reports that Aristotle and other academics distinguish problems, the geometrical propositions that involved action and production procedures—including construction, division into sections, and subtraction and addition of parts—from theorems, those propositions that cannot be described without action language. As Proclus puts it, "problems announce that something is to be done, theorems that some truth is to be discovered and demonstrated" (*In Eucl.* 210). Thus, the philosophical question of how to understand "changes" in geometrical objects corresponds to a methodological debate concerning the status of construction in scientific geometry.

Followers of Speusippus and Amphinomus insist that only theorems be included in mathematics, which they conceive as a theoretical science of unchanging things (*In Eucl.* 77). According to Amphinomus's taxonomical scheme, geometrical propositions with one solution are "ordered" (συνιστάμενα), those with more than one but a finite number of solutions are "intermediate" (μέσα), and those with infinite solutions are "unordered" (ἄτακτα) (*InEucl.* 220). Like Aristotle, Amphinomus is interested primarily in the logic of geometry.

Proclus attributes the distinction between problem and theorem to Aristotle, and asserts that Amphinomus adopted the view that geometry does not investigate causes from Aristotle (*InEucl.* 202).[2] I am unaware of any passage in which Aristotle makes this claim. Yet in the *Eudemian Ethics* (1222b24), Aristotle writes that a "principle" of unchanging things, such as in mathematics, is not meant in the proper (κύριος) sense of the word. Yet a principle could only bring about a "change" in mathematics by destroying a hypothesis. Human agents, in contrast, are principles in the proper sense that they bring about the motions that constitute action. Aristotle's point seems to be that principles of unchanging things cannot be causally efficacious: they only bring about "changes" in an argument, not in the objects of mathematics themselves.

Lachterman argues that there is a passage in which Aristotle explicitly addresses the puzzle of change in geometry, corroborating Proclus's claim that he sided with Amphinomus and Speusippus. In *On the Heavens*, Aristotle objects to those Platonists who claim that the world is indestructible but nevertheless offer a description of its genesis. These philosophers hold that this kind of exposition is analogous to that of geometry. "In the production (ποίησις) of diagrams, all the initial components having been set out, the same comes about simultaneously. But in their demonstration, the same does not come about simultaneously. But this is impossible. For the things assumed earlier and subsequently are contradictory. They say that ordered things came to be out of disordered, but it is impossible for the same to be simultaneously disordered and ordered. There must be a genesis involving the separation of things in time, as well. In the diagrams, nothing is separated in time" (*De Caelo* I.10, 280a4–11).[3]

Aristotle argues that in setting out a geometrical construction, there can be no contradiction among any of the components. Consequently, though a geometrical construction might be presented as a generative process, every step in this genesis really occurs "at the same time." In contrast, physical generation involves real change and thus cannot be simultaneous. For example, in Plato's creation myth in *Timaeus* (30a), God brings order out of disorder. But nothing can be at once ordered and disordered: these heterogenous components can only coexist over time. The components of mathematics, however, are homogenous. Consequently, mathematical genesis, for Aristotle, as for Amphinomus and Speusippus, "has an 'as-if' character."[4] These thinkers suggest that all geometrical propositions should be interpreted as theorems: geometrical constructions are not productions but acts of understanding.

Opposed to Speusippus, the school of Menaechmus holds that all propositions are problems, dividing them into those meant to produce something sought after and those that involve seeing some feature of a construction (*InEucl.* 77). Menaechmus, like Aristotle, was a follower of Eudoxus, a student of Plato, and aimed to "perfect" geometry (*InEucl.* 67). Menaechmus thus was both a leading geometer, contributing especially to the problem of cube duplication,[5] and worked in logic, on the relations among propositions. For Menaechmus, elements are not just the basic building blocks of propositions, like letters that make up a text, but can signify completed constructions implicit in problems that come later in the sequential development of a proposition (*InEucl.* 72–73). Before the codification of geometry into a system of elements by Euclid, Menaechmus saw that by making all propositions problems, basic constructions could be called on recursively in further constructions. Whereas Plato advocates for an ascent from hypotheses to dialectical definitions in the *Republic*, Menaechmus takes the deductive path of "descent" from elements to ever more complex constructions to be the guiding principle in the development of a scientific geometry.

This debate between "formalists" and "constructivists" exemplifies how Plato's written works were discussed and recast as technical questions within the Academy. While the treatment of geometrical production in the *Republic* is humorous and artful, the philosophical points made in support of each side of the Academic debate are informed by the technical questions of the day. On the side of the formalists, concerns about the metaphysical status of intermediates as eternal and unmoving beings, as well as a preference for geometrical theorems with unique or at least finite solutions, drove a view on which all propositions are theorems. On the side of the constructivists, a desire for geometrical constructions that could ensure additional steps in a progressive sequence, and an attitude toward categorizing problems in terms of the activity of the geometer rather than the ontology of the objects constructed, motivates a perspective from which action vocabulary, far from seeming ridiculous, gives clues about how to direct geometrical research. The platonic puzzle about the "doings" of mathematicians, like any good philosophical puzzle, is not only an entertaining vignette meant to inspire curiosity but also provides a point for debate between two different attitudes toward the science of geometry.

Plato's paradox can be observed directly in theorems or problems that use the productive language of kinesthetic construction. According to Proclus, every geometrical proposition has parts that are arranged in a specific order. The πρότασις puts forth what is given and what is sought. The ἔκθεσις sets out what is given. The διορισμός specifies what is to be constructed. And the κατασκευή, or "construction," completes the figure. After these constructive steps, the ἀπόδειξις, or demonstration, proves the proposition from the diagram. Finally, the συμπέρασμα, or conclusion, restates the general results (*InEucl.* 203, 209–210). Whether this schema was already in use in Aristotle's or Euclid's time, or whether it represents a later reconstruction of Euclidean technique, is an open question.[6] The salient point here is just that, in antiquity, at least some philosophers took it as obvious that deductive proofs could only be made after the completion of constructive steps expressed by motion verbs. Though it is a theorem rather than a construction problem, proposition 37 of the first book of Euclid's *Elements* provides a dramatic illustration of this phenomenon.

The proposition begins with the annunciation: to show that two triangles of the same base and same height are equivalent in area. It then sets out what is given: on the left side of figure 1, suppose that there are two given triangles, ABC and DBC, on the same base BC, which is parallel to AD. What must be shown is that these two triangles are equal. The construction that follows transforms the figure on the left to that on the right. Let AE be produced in both directions to arbitrary points E and F. Let a line be drawn through B to E. Let a line be drawn through C to F. These steps make use of the motion verbs that Plato held were ridiculous but ineliminable from the geometer's procedure. Euclid uses perfect, imperative verbal forms for "produce" (ἐκβεβλήσθω) to extend

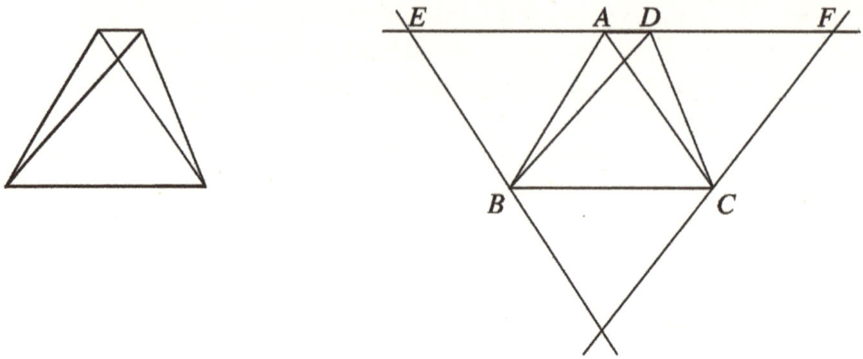

FIGURE 1

the parallel AD in both directions, and of "draw" (ἤχθω) to indicate the penetration of the line parallel to AC through B, and the line parallel to BD through C. As imperatives, these verbal forms tell the geometer what to do to produce the figure. The construction thus appears to be a productive process that terminates in a finished diagram that is presented to the geometer as available for further demonstrative reasoning.

The demonstration that follows this constructive part of the proposition consists of a series of assertions, which are stated as consequents of the construction, then the parallelograms EBCA and DBCF, since they are on the same base BC and the same parallels BC and EF. Moreover, the triangle ABC is half of EBCA, for the diameter bisects it, and the triangle DBC is half of the parallelogram DBCF, for the diameter DC bisects it. The final step is to use a general principle (Euclid's third common notion) that the halves of equal things are equal to each other. The conclusion follows that the triangle ABC is equal to the triangle DBC. Since demonstrations are deductive, they must be stated in a series of sentences.

Understanding this discursive articulation of geometrical reasoning depends on matching each proposition to a part of the construction. Thus, there is a representational aspect of the geometrical proposition distinct from the demonstration but upon which it depends. Not only is there a requirement for productive change in making the construction but also a geometrical "vision" of the product of that construction is needed. Geometrical production terminates in geometrical presentation.

Although Plato insists that geometry studies being rather than becoming, it is clear that geometrical objects could not be the forms themselves, grasped by intellect. The example considered here begins with two triangles rather than a singular form. But discursive thinking alone cannot account for the geometrical procedure. For although geometrical demonstrations are discursive, they depend on the construction of geometrical figures, the parts of which can be moved and arranged prior to the demonstration. Grasping the geometrical proposition depends on both a production and presentation of diverse objects. What power of the psyche could produce and apprehend objects that are not fully articulable by discursive reason but that could also never be directly perceived?

Division (διαίρεσις) and collection (συναγωγή) are terms used throughout Plato's works, and their precise meaning may vary by context. In the *Phaedrus* (265e–266c1), Plato recommends them as proper methods of dialectic, which distinguishes elements by cutting up (διατέμνειν) a subject matter according to some criterion.[7] Division and collection are supposed to be constitutive of any body of systematic knowledge. Jaeger argues that Plato's introduction of the method was necessitated by a demand for classification of ideas.[8] In distinction, Proclus reports that the method of division originated in geometry, in which it was used to divide figures, and that Plato recommended its use "in all the sciences" (*InEucl.* 211–212). Whatever its initial role in Plato's philosophy, the method of division remained a standard logical topic in the Academy even after Plato's death.[9]

It is unsurprising, then, that division is the method used in the *Republic* to distinguish powers of the psyche. "In a power I see neither some color nor shape nor anything else of the many others, which by observing in other cases I can distinguish for myself some thing from others. Rather, in this case I look solely to what it is about and what it accomplishes, and on account of this I call each the power that it is. And that which is about the same thing and which accomplishes the same thing, I call the same, and that which is about something else and accomplishes something else I call a different power" (*Pol.* 477c–d). Powers of the psyche are to be distinguished from one another by the things they grasp and by the things they produce. While these two criteria are conceptually distinct, Plato takes them to be inseparable. Thus, each power must be set over a unique class of objects, to which it gives its possessor exclusive access.[10]

Though Aristotle often criticizes Plato and his followers on methodological matters, he evidently adopts this criterion of psychological division. Hence while he never definitively answers his opening question of whether division could serve as the general method for psychology (DA 402a19–2), *De Anima* consists largely in dividing capacities based on distinctions between proper, common, and incidental objects. Likewise, in the *Ethics* he argues on the hypothesis that knowledge depends on "a likeness and kinship" between the knower and object known, that different genera of things can be said to correspond to distinct parts of the psyche (EN 1139a9–12). Thus, intellect and perception can be "cut off" from each other due to their having different objects (εἰς τὰ πράγματα), presumably universals and particulars. This difference in objects requires a corresponding division between powers, since it is in virtue of their objects being in potential or actualized (εἰς τὰ ἐντελεχείᾳ) that parts of the psyche are in potential or actualized (DA 431b24–26).

Though they are different in potential, in an actualized state of intellect or perception, both a power of the psyche and its object are mutually activated or exercised (or, as Aristotle puts it in a passage in the *Metaphysics* discussed in chapter 5, "actuated"). A stinky cheese is not a psychological power, and the sense of smell is not a dairy product. They are different in potential. But it is only when a cheese stinks and an animal smells it that the cheese (qua stinky) and smelling both come to be what they really are. And it is because cheese is stinky, and thus apt to be smelled, and because the sense of smell is a power exercised by the nose, and thus apt to smell nearby stinky cheese, that the activity of smelling a stinky cheese is possible. The generic match between object and power—being the kind of thing to be smelled and being the kind of thing to smell—is thus explanatory of their mutually constitutive activity. Consequently, generic differences among objects is Aristotle's criterion for psychological division.

There is, however, considerable disagreement about how to interpret Aristotle's division of the psyche. Johansen holds that Aristotle is "the progenitor of faculty psychology," since he divides powers asymmetrically among various forms of life.[11] Weinman, perhaps following Lear's characterization of Aristotle as a "holist,"[12] holds that the parts of the psyche for Aristotle are "taxonomical rather than ontological."[13]

This question of whether the parts of the psyche are *de re* or merely *de dicto* divided is not obviously answered in Aristotle's corpus. Aristotle sometimes writes as if psychological parts are really distinct—for example, dividing the psyche hierarchically into nutritive, perceptive, desiderative, locomotive, and thinking powers (DA 414a31–32). In that case, he takes the lower power to be "in" the higher, just as the triangle is "in" the square (DA 414b29–32, cf. 20–22). Any square can be divided into two triangles by drawing a line diagonally from a corner. But a triangle cannot be divided into squares. Analogously, any animal that can, for example, taste will also have the nutritive power by which it takes nutrients from its environment. But tasting will not be found in every animal that has the nutritive power. The triangle and the nutritive power in this sense are "in" the square and tasting, though the converse is not true. Such necessary orderings, in which the prior thing constitutes the posterior thing in the sequence, suggests that Aristotle thinks of the divisions among powers of the psyche as being just as real as the distinction between triangles and squares.

Elsewhere, however, Aristotle claims that distinctions among parts of the psyche are merely nominal. Thus, all the parts of the psyche "are of the same form (ὁμοειδῆ) as each other and as the whole. They are like each other because they are not separable (οὐ χωριστὰ) and like the whole in that it is not divisible in substance (οὐ διαιρετῇ οὔσῃ)" (DA 411b24–28). If the parts of the psyche are indivisible in substance, then it seems that psychological divisions are logical rather than ontological. If they are inseparable from one another and from the whole, then tasting and taking nutrition from the external environment, for example, may just be different descriptions of the same activity. Thus, "exercising the nutritive power" and "exercising the power of taste" might be two different descriptions of a lion eating a piece of meat, rather than two really distinct psychological activities.

The embattled question of whether psychological parts for Aristotle are de re distinct thus proves to be difficult. Happily, Aristotle is at least clear that parts of the psyche are de dicto distinct, in the same way the concave and the convex are distinct in account (EE 1219b33–38, cf. EN 1102a126–32). Here the analogy is comparatively clear. A seashell, for example, will only be correctly described as concave when it is also correctly described as convex. But only descriptions of the seashell that refer to its concavity will be useful for explaining why, when one puts it to one's ear, one hears the ocean. Descriptions of its convexity, in this context, are superfluous. Even though they are not really distinct, dividing concavity and convexity is still scientifically useful, isolating the relevant descriptions for explanation of a certain phenomenon.

The method of division, by which one "cuts off" a power that grasps a distinct intentional object, de dicto separating it from the others by giving it a distinct description, would have struck any Academic philosopher as the appropriate method for explaining cognition of that kind of object. For example, one can see and hear a stormy ocean. But its frothy green color can only be grasped by sight, its roaring only by hearing. The existence of such propria is a primary way in which Aristotle divides the senses as powers (DA 418a7–20). The fact that these powers have common objects—motion, and so on—is, moreover, what makes them collectively as the complete power of perception. Given this way of dividing and collecting psychological parts in terms of their objects, there is nothing innovative about employing the method to explain access to geometrical objects. Nor is it remarkable that Aristotle explains the cognitions underlying geometrical construction by describing a psychological power capable of production and presentation. What needs to be explained is why Aristotle identified this power as φαντασία, which is to say appearance, and how he invented imagination in the process.

A great deal has been written about Aristotle's characterization of imagination, but the original motivation of Aristotle's invention of the new power has remained obscure. Since "imagination" was never used as a technical term before Aristotle, what leads him to define it and what sense does it have for him? The problem faced by Aristotle and the generation of philosophers at the Academy after Plato was twofold.

First, it is required that there be a capacity for constructing intermediate, intelligible figures that were nevertheless extended and diverse. This is necessary for all geometrical proofs requiring postulates, which allowed figures to be kinesthetically constructed before discursive inferences were made about them. With the introduction of an axiomatic structure to geometry that allowed for progressive series of demonstrations in which later propositions could follow and make use of prior propositions, a further extension of this production was required. Not only should "self-evident" postulates be utilized in constructions, but also "established" productions should be able to be called on recursively to produce new figures. It is clear at least that the development of scientific geometry utilizes a kind of formal "memory" in each repetition of an established production. Unlike the imperfect memories of perceived objects, this memory of intelligibles should invariably produce what was demonstrated before within the new construction. A productive power was required that could both reproduce prior propositions "from memory" and constitute a potentially infinite number of new figures spontaneously.

Second, a state of psyche was required that could present an intelligible figure to discursive reasoning. This presentation could not itself be productive, since its production might be achieved stepwise, while the completed construction must be presented all at once. Nevertheless, this presentation has to display the achievement of the activity of construction and reinstantiate the prior results in the axiomatic progression. For example, in the Euclidean proposition rehearsed in figure 1, one concludes that two triangles on the same base are equivalent by applying the rule that halves of equals are themselves equal. In order to apply this rule, the geometer must "see" that the distal points lie on the parallel, and that the constructed parallelograms are themselves equal. Discursive reasoning itself could neither produce geometrical constructions nor present the completed constructions to itself.

Aristotle's procedure in dealing with this problem is first to show that there is a power distinct from perception (DA 427b27–28) but that nevertheless has perceptual features. Thus, Aristotle divides perception from thinking before turning to give an account of imagination in relation to the other psychological powers and states (DA 428a1). Aristotle adopts the nonpsychological term "phantasia" from Plato's writings but gives it a new technical function in his systematic psychology that is both productive in the sense of the geometrical construction and presents something absent, the perfect geometrical figures suggested by the perceptible diagram. Aristotle then separates imagination from perception, showing that imagination is a kind of thinking, and from belief, showing that the new power is distinct both from discursive thinking and from intellect. At the end of his technical treatment, Aristotle offers a definition of phantasia and suggests that the term derives from "*phos*" (light), implying that it is best understood as a kind of productive vision of intelligible objects. This structure of the third chapter of the third book of *De Anima* comes clearly into view when one appreciates how Plato's paradox of geometrical alteration served as the guiding thread that Aristotle followed in his investigation.

Aristotle adopts the term "phantasia" from Plato. Silverman notes seven uses of the word in the Platonic corpus, arguing that imagination plays a major role in Plato's psychology, as the mental power that links nonpropositional perceptual states with truth-bearing cognitive states.[14] Schofield and Albinus or Alkinoos, author of the middle-platonic *Handbook*, likewise take Platonic phantasia to indicate something in the psyche.[15] Aristotle treats these characterizations as endoxa, the hypotheses by which he establishes his own division. Yet inspection of these passages shows that Plato nowhere uses "phantasia" to describe a part of the psyche.

Plato uses the word "phantasia" in the *Republic* in the context of an argument meant to establish, against Homer and the tragedians, that a god could not be a deceiver (*Pol.* 382e). This is so, Socrates claims, because demons and divinities are true both in deed and in word. Hence insofar as he is good and truthful, god could not deceive in appearances (φαντασίας), words, or signs. As O'Connor notes, this comment highlights the difference between gods and men since it is only the latter who tell lies and disguise themselves from enemies.[16] This use of "phantasia" critiques poets who portray gods in human attitudes, and thus as being less than completely true in word and deed. Socrates attributes such flawed portrayals of the gods to Homer, who describes Zeus sending a deceiving dream to Agamemnon (*Il.*II:1–35) and Aeschylus, who makes Apollo lie to Thetis at her wedding (*Pol.* 383b). The passage focuses on the deceitful behavior on the part of the poets' gods, with the term "phantasia" used to indicate an appearance that could be either true or false. This appearance is, however, not something psychological but rather an act of deception, an explicit word or deed that intentionally misleads its audience. In the *Republic*, "phantasia" means not a power of the psyche but one's costume or behavior.

Plato uses the term "phantasia" again in the *Theaetetus* in his description of Protagoras's doctrine that "man is the measure." According to Plato, Protagoras's view that everything "is to me such as it appears to me, and is to you such as it appears to you" entails that the perception that something is hot or cold is always veridical (*Theaet.* 152a). But if this were the case, then to appear (φαίνεται) would be the same as to perceive, since something can appear hot to one person and cold to another. It follows that phantasia is the same thing as perception (*Theaet.* 152c1). "Phantasia" here is the substantive of the verb *phainetai*, an "appearance" in general, rather than a special power.[17] Thus, a little later in the same dialogue, Plato employs the term "*phasmata*," which perhaps should be corrected to "phantasmata," to indicate "apparitions" or "phantasms" (*Theaet.* 155a). Although it is retrospectively possible to read "phantasia" as referring to a power, and "phantasmata" as referring to psychological states, internal to the dialogue, there is no suggestion that the terms should be understood in a psychological sense. Plato uses these words naively rather than technically, as general appearance-terms that help describe and refute the false theory of Protagoras.

A third appearance of the term in Plato's corpus occurs in the *Sophist* (264a–b), in the Stranger's description of perceptual judgments. When one makes a judgment based on a perception, what it produces ought to be called "phantasia." Thus, when we say that something appears, we indicate a combination of perception and belief (σύμμειξις αἰσθήσεως καὶ δόξης). Of all the instances of the term in Plato, this seems most like a psychological definition of phantasia. Aristotle takes this as a potential definition and offers a vigorous argument against the doctrine that phantasia is a combination of perception and belief (DA 428a16–b9). But Plato here considers phantasia as a product of perception, which is wrongly endorsed by belief. Thus, phantasia here appears to be a state rather than a power of the psyche. Yet on Plato's view that different powers have different objects, it seems that no such mixed state could exist. Thus, Plato may be using the term "phantasia" negatively, to show that perceptions are insufficient to justify discursive judgments.

I have been arguing that Plato did not attribute any psychological significance to "phantasia" but used the term as a substantive correspondence to the verb *phainetai*. Indeed, if Plato's descriptions were intended as definitions of phantasia, they would be inconsistent. Phantasia cannot be both perception and a mixture of perception and belief. Aristotle's treatment of Plato's naive characterizations as candidate definitions of a psychological power and the division he makes is not engagement with views actually propounded in the dialogues but rather with potential definitions available to his interlocutors in the Academy. Aristotle's division is thus not in disagreement with Plato but with anyone who would interpret Plato's naive statements as definitions of phantasia.

CHAPTER 3

IMAGINATION DEFINED

De Anima III.3, Aristotle's central chapter on the imagination, begins by distinguishing perception from thinking (Aristotle's argument is interesting in its own right, and is discussed at greater length in chapter 4). He hypothesizes that there are two kinds of thinking, phantasia and judgment (ὑπολήψεως), the latter of which he takes to include science, belief, and practical reason (ἐπιστήμη καὶ δόξα καὶ φρόνησις). He then offers reasons for dividing phantasia from perception, and from judgment (DA 427b27–28). In making these divisions, Aristotle uses the two opposed descriptions of phantasia from Plato's dialogues as *endoxa*, according to which his own divisions can be made. Aristotle divides phantasia from perception with six arguments.

1. "Perception" means either a power in potentiality (δύναμις) or an actualized capacity (ἐνέργεια), e.g. the power of seeing (ὄψις) and sight (ὅρασις). But something appears (φαίνεται δέ τι) even when neither potentiality nor actuality is asserted (μηδετέρου ὑπάρχοντος τούτων) of perception, such as in sleep (DA 428a6–8).

2. Perception is always present (ἀεὶ πάρεστι) but phantasia is not always present (DA 428a8–9).
3. If actualized phantasia (τῇ ἐνεργείᾳ) and actualized perception were the same, then phantasia would be in all beasts (ἐνδέχοιτο τοῖς θηρίοις) but phantasia is attributed only to some animals, such as the bee and ant, but not to other animals, such as the grub (DA 428a9–11).
4. Perception is always veridical (ἀληθεῖς ἀεί) but phantasia often is false (αἱ δὲ φαντασίαι γίνονται αἱ πλείους ψευδεῖς) (DA 428a11–12).
5. It is not when we are activated accurately with regard to the object of perception (ὅταν ἐνεργῶμεν ἀκριβῶς περὶ τὸ αἰσθητόν) that we say that "*this* appears" (φαίνεται τοῦτο) to us to be a man but rather when we are not actively (μὴ ἐναργῶς) perceiving whether it is true or false (DA 428a12–15).
6. Things appear also to those whose eyes are shut (DA 428a15–16).

These arguments presuppose that perception is always of a present object. Thus, in arguments 1 and 3, Aristotle asserts that even when perception is not activated by a presence, it is possible for something to appear for us. These appearances, or phantasmata, need not be present before the senses in order for one to "see" them (argument 2), are not always true of the thing they are about (argument 4), and are made present when one is not actually perceiving (argument 5) or even when one is not perceiving at all, such as when one is unconscious (arguments 6). These arguments show that perception grasps veridically what is present to the senses, while phantasia grasps absent objects and thus presents both true and false appearances.

One might think that phantasia is belief, since unlike science and intellect, which are always true, belief becomes both true and false (DA 428a16–19). Aristotle observes here that if his characterization of phantasia as a distinct part of the psyche is correct, then he must also be able to divide phantasia from belief. He argues:

7. In believing something one has conviction (πίστις) but imagining something does not produce conviction (DA 428a19–21).
8. Conviction is not attributed to beasts but phantasia is attributed to many beasts (DA 428a21–22).[1]
9. If imagination were belief, then for x to appear to one as φ, one would believe that φx. But something appears to one as φ without one's believing that it is φ (DA 428a24–428b2).

10. Something can appear falsely even when one has a true judgment, such as when the sun appears to be the length of a foot, while one is convinced (πιστεύεται) that it is greater in size than the inhabited world (DA 428b2–4).
11. Suppose that one believes φx and that φx is true. If imagination is belief, then when one imagines ¬φx, either
    i. One has changed one's belief from φx to ¬φx, without any change in the facts, without having forgotten anything, and without being persuaded to the contrary; or
    ii. One believes something true and something false, viz. that "φx & ¬φx."

    In either case, one would be able to change one's true belief without being presented with any evidence contrary to that belief.

Aristotle uses these five arguments to refute both the direct identification of phantasia with belief and the combinationist view suggested by Plato in the *Sophist*. The digressions in the division (DA 428a22–428b1) confusingly interwoven with arguments 8 and 9 are meant to show that these arguments work against both the direct identification with belief and against the combinationist view. Arguments 7, 10, and 11 speak against both views, justifying Aristotle's conclusion that phantasia is neither perception nor belief, nor a combination of perception and belief (DA 428b9). The conclusion of all these arguments is that believing a proposition involves a commitment to the truth of that proposition, while enjoying an appearance requires no such commitment.

To Aristotle's arguments for the division of other psychological powers from phantasia may be added two arguments given earlier in the chapter as introductory to the division. These arguments give the deeper reasons why phantasia must be individuated through the division:

12. As opposed to all other thinking faculties, we are affected by phantasia whenever we wish. For phantasia produces (τι ποιήσασθαι) something before our eyes (πρὸ ὀμμάτων), like what is placed (τιθέμενοι) for those who make images arranged in mnemonics (DA 427b17–19).
13. Whenever we believe something is terrible and fearsome we immediately sympathize with it (εὐθὺς συμπάσχομεν), and similarly with what is comforting. But we hold to phantasia as if (ὥσπερ) we were looking at a terrible or comforting picture (θεώμενοι ἐν γραφῇ τὰ δεινὰ ἢ θαρραλέα) (DA 427b21–24).

These last two arguments must be given a privileged place. For argument 12 entails that phantasia can be productive in the way required in making geometrical constructions, while argument 13 makes phantasia a kind of nonperceptual presentation of absent things, which is just what is required for the grasp of intermediate figures.

Phantasia, in argument 12, is taken to be productive in the same sense as an art that produces material for use by someone else.[2] Indeed, I argue in the following chapters that the states produced by phantasia—the phantasmata—are used by the intellect, both in theoretical endeavors such as geometry and in practical deliberation. Aristotle here attributes to phantasia a distinct, presentational function when he says it places its productions—the phantasmata—"before our eyes." Thus, we behold phantasmata as if we are "looking at a picture" (13).[3] Aristotle uses the same locution in the *Poetics* (1455a22–26), where he claims that a good dramatist will use plot and diction to produce the most vivid scene possible before his audience's eyes. As in the *Poetics*, the phrase "before our eyes" denotes an act of display. And, in argument 13, it turns out that this display is distinctive in that it does not command an emotional response, as do changes in perception and belief. When phantasia presents one with a scene, one is typically, though not invariably, aware that things may not really be that way. Phantasmata have a sense of unreality, comparable to looking at a painted image. Aristotle's comments suggest that phantasia is a power—the imagination—that is a part of the psyche distinct both from the lower perceptual powers and from higher forms of judgment.

Aristotle offers three candidate definitions of phantasia in *De Anima*. One is "phantasia seems to be some movement that comes to be not without perception" (DA 428b11–12). According to this description, perception may be moved without phantasia, though the converse is not true, since every animal that has phantasia also has perception. It follows that the motion of phantasia cannot exist in an animal unless that animal also has a perceptual capacity. Yet according to arguments 3 and 5, the activity of phantasia and of perception are not the same. In distinction to perception, which can be understood as the reception of a perceptual form from the thing perceived, the activity of phantasia involves an active production of an image. While this description gives the specific differences between phantasia and other forms of thinking that Aristotle groups under judgment, it does not differentiate the appearances produced by phantasia from the afterimages produced by perception. Thus, it is not a definition of imagination.

The word "phantasia" can be used in a metaphorical sense to indicate anything that is merely apparent. But in its technical meaning, Aristotle says, "phantasia is that in virtue of which we assert that some phantasm comes to be (γίγνεσθαι) for us" (DA 428a2–4). This statement cannot be a definition, since it describes the power in terms of the states generated by that power. However, when phantasmata are understood as distinct from the afterimages generated by perceptual activity and the objects of understanding grasped in intellectual activity, it is clear that this statement picks out a distinct power. This formulation uses the productive vocabulary of argument 12, saying that a *phantasma* are generated or "come to be." Likewise, the phrase "for us" in this definition can be interpreted via argument 13 as showing the presentational aspect of phantasia. In characterizing phantasia as the producer of nonperceptual phantasmata that one is able to "see" or "envision," it is clear how imagination fills the roles required for it in geometry: the imagination is a distinct power of thinking in virtue of which images of absent things are made present to us.

The third statement, which must be Aristotle's definition, is that "phantasia would be a movement, which comes to be in virtue of the actualization of perception" (DA 429a1–2). Phantasia cannot exist in an animal without there also being a perceptual capacity that becomes actualized. Imagination thus conceived is a psychological motion or change, which is consistent with the productive and presentational functions employed in Aristotle's division. The problem of Aristotelian imagination is how to give a coherent characterization of the imagination as a distinctive power of the psyche that satisfies all three descriptions.

The close association of imagination with perception in Aristotle's definition has led many commentators to understand imagination as a queer type of perception, or as an aftereffect of normal perceptual experience. Schofield holds that Aristotle understands the imagination as a capacity for having "non-paradigmatic sensory experiences."[4] Nussbaum argues that Aristotle primarily understands the imagination as a "seeing as."[5] In this view, perception renders uninterpreted information, for example, sounds and colors, and imagination provides the "interpretive side" of perception, in virtue of which one perceives, for example, the sounds as meaningful words or the colors as the surface of a solid object.[6] The challenge for this interpretation is to explain why, if its primary function is to contribute to perceptual experience, Aristotle needs the imagination at all. Others take the view that the imagination in these passages is not a form of perception but, as Sorabji puts it, "the power of having images."[7] In some versions of this view, the phantasmata are understood as being "like the objects of sense except for being wholly internal."[8] The challenge for such readings will be to explain how the power of imagination can somehow be the same as the power of perception, while the essence of each faculty is different (*Insom.* 459a16–21). One strategy for this approach is to say that imagination and perception share the same content, a claim to which I return later on, though they put one in a different relation to that content.[9] This sort of view tends to conceive of the imagination as the retaining of information that is originally encountered in the perception of a present external object, after that object has become absent.[10]

Both interpretations get something right about Aristotle's doctrine of the imagination, and both risk misunderstandings. Iconoclasts who take the imagination to be a modification of sensation or queer sort of perception get right the idea that Aristotelian imagination can pick up on minute features of the objects given in perceptual experience, rearranging and extending them in novel ways. But they tend to gloss over Aristotle's descriptions of the imagination as the productive and presentational presencing of absence. The imagination is a motion of the psyche that produces and presents one with a phantasma, which is not present before the senses. But there is no suggestion that the absent thing must be something one has actually previously encountered in perception. A phantasm might represent what one saw yesterday but which is now absent, or it might present a foot the size of the sun, which is a physical impossibility and thus completely unreal.

Iconophiles get right the productive and presentational or depictive aspect of Aristotle's descriptions of imagination. But conceiving of phantasmata as mental *images* takes the focus off Aristotle's dynamic characterization of the imagination in his definition. Imagination is a *kinesis*—a motion or change—and there is no suggestion that the phantasma produced in that motion must be a static image. Indeed, Aristotle typically treats the phantasm as representing "unreal" motions in things, as required, for example, for extending the parallel in the geometrical example considered earlier. The phantasm "animates" the static geometrical diagram, rather than reproducing it in an internal copy. Second, conceiving of phantasmata as *mental* images risks introducing a distinction, foreign to Aristotle's psychology, between psychological interiority and exteriority. To be sure, Aristotle conceives of the imagination as power of motion in the psyche, and its phantasmata as psychological items or events. And in this sense, they may be subjective and private. But they must not be taken to inhabit an internal "place" of representations. For the psyche is not an immaterial mind, or introspectable consciousness, but the form in virtue of which an animal lives.

Aristotle's definition is commonly taken both by iconophiles and iconoclasts to suggest that imagination merely reproduces perceptual experiences. In this view, the content of a phantasma is just the content of the actualized perceptual power, when it is retained in the animal.[11] The definition is thus taken to be an efficient causal account of the activity of imagination that makes its content completely dependent on the immediately preceding perception. But if Aristotle meant to indicate that imagination is merely reproductive of perception, one should expect him to say "phantasmata" here rather than phantasia—that is, to indicate a state-state parasitism of imagination upon perception. Far from asserting this, Aristotle's definition merely establishes that imagination is a change in the psyche set in motion by perception. This does not foreclose on the possibility of its producing something not originally present in that perception.

If the "reproductive" conception of imagination is not supported by Aristotle's definition of imagination, it is contradicted outright in other passages. In *On the Motion of Animals*, for example, he holds that "imagination comes about either through thought or through perception" (γίνεται ἢ διὰ νοήσεως ἢ δι'αἰσθήσεως) (MA 702a19). This entails, I think, that on some occasions, the motion that is the imagination is set off by perception and, on other occasions, by the intellect. In cases in which it is triggered by perception, I shall argue, the imagined state of affairs bears a similarity to the perceptual episode that triggered it. But in cases in which it is triggered by the intellect, there appears to be no such requirement. In both cases, however, imagining produces an arrangement of forms in a phantasm, akin to a mnemonic, and presents them to one in a way that is like looking at a picture.

The capacity of the intellect to determine the imagination is not very surprising. For Aristotle's primary purpose in dividing the imagination from perception and belief and defining it—the one he uses in the sections of *De Anima* immediately following the division—is to explain how the partless objects of intellect are thought as articulated numerically or mereologically and thus as subject to operations that "change" them. Among these operations are, presumably, the cognitive acts of division, duplication, and augmentation necessary for extending, drawing, or producing a geometrical figure. The postulates required for the kinesthetic construction of geometrical objects thus find their cognitive ground in the imagination. Aristotle writes that the psyche never thinks without a phantasma (οὐδέποτε νοεῖ ἄνευ φαντάσματος) (DA 431a16–17) and that the power of thought thinks intellectual objects in phantasma (εἴδη τὸ νοητικὸν ἐν τοῖς φαντάσμασι νοεῖ) (DA 431b2). For instance, mathematical objects are thought as existing separately, although when they

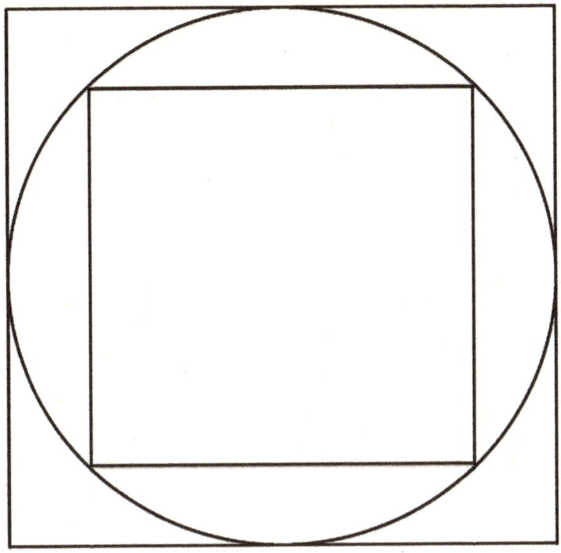

FIGURE 2

are being thought, they do not exist separately (DA 431b15–17). Geometrical objects that are separable from the matter of the diagrams in which they are represented can be thought apart from the perceptible diagrams. A figure is thus "alienated" from its perceptible representation in a diagram, re-presented in the imagination, but grasped veridically as an intellectual form.

This doctrine of the imagination allows Aristotle to treat mathematical objects as objects of intellect, despite their representation in a geometrical figure. In a passage I discuss in greater detail when I come to his account of geometrical cognition, Aristotle argues that because no thinking occurs without an image, thinking involves a kind of "drawing" in the imagination. In this case, something without a quantity is represented as having a definite quantity (*Mem.* 449b30–450a5). Suppose one thinks of an intelligible form that does not itself have a quantity. In order to think it, one must represent it in a phantasma "before one's eyes." In thinking of a circle and a square, for example, one must draw them out, either on paper or in one's imagination. Once they have been drawn out, however, the two shapes come to stand in a definite relation, for example, as a circle that passes through the corners of the square (figure 2). In representing them in this way, one imagines a figure in which the circle is of a greater size than the square, putting the shapes before one's eyes as definite quantities that can be compared to one another. Yet these phantasms cannot be identical to the concept (νόημα) of the circle or square, which do not have a definite quantity. The figures might be drawn differently, with the circle as touching each side of the square, so that the square is larger than the circle. But the concepts would remain unchanged. Aristotle's point is that thinking of ideal circles and squares, which have no definite quantity, depends on picturing them *as* quantities—that is, as extended figures drawn by the imagination. The division of the imagination from the other powers of the psyche thus accounts for how absent objects can be produced and presented in particular configurations.

If the theory of the imagination was originally formulated by Aristotle to accomplish the peculiar tasks set by scientific geometry, the concept of appearance is nevertheless transformed in his psychology. Whereas for Plato and the preceding tradition, a phantasia or phantasma was an appearance produced by a perceptible object in one's nearby environment, for Aristotle, these are technical terms indicating a psychological power and what is produced and presented by that power. Aristotle's fundamental move is to forge the substantive concept of appearance into a part of the psyche in virtue of which one enjoys appearances, when one is set in motion by perception or by intellect. The phantasmata, which are presentations of absent things, are irreducible to the general pre-Aristotelian appearances, which were invariably caused by something present in the world of sensation.

If this reconstruction of the original meaning and purpose of the imagination is accurate, then the concept of imagination should be considered an innovation of theoretical philosophy. This innovation is motivated by the problem of why knowledge of atemporal must be grasped through a temporal process of generation. Aristotle's answer is there is a generative motion that accompanies thinking the universal concepts. But this motion is psychological and merely apparent—it is the imagination, rather than the reality. Whereas Plato had assumed that the power of appearing was a property of things, Aristotle makes it a part of the psyche. Yet no sooner is this power of producing and presenting mere appearances transferred from the world to the psyche than it becomes an irresistible resource for describing a wide variety of psychological phenomena. Indeed, this resource is so powerful that one is sometimes apt to forget that the imagination is not something natural or obvious but an invention designed to answer a Platonic riddle about mathematics.

CHAPTER 4

DREAMING *and* DAY-DREAMING

The third book of *De Anima* forms Aristotle's systematic account of the human psyche. Beginning with a recapitulation of the perceptual theory applicable to all forms of animal life developed earlier in the treatise, the latter chapters turn to the consideration of the peculiarly human thinking before outlining a psychology of action. The third chapter of that book is pivotal in introducing the imagination and marks a transition from animal perceptual powers to the discussion of the human intellect. What explanatory roles does this new power play in Aristotle's psychology?

Aristotle begins the chapter by attacking the doctrine of "the ancients," that the rational psyche and its objects are embodied like sensory perception (αἴσθησις). In his work *On Nature*, Empedocles claimed that "the intelligence of man grows towards that which is present (πρὸς παρεὸν)."[1] The presence of the object of cognition to the organ of cognition is a necessary condition for cognition. Empedocles takes the heart to be the central organ of cognition, which is connected to the peripheral organs through the bloodstream. His description of this communication of the blood to the heart supports a physicalistic theory of mind in which "the blood round the heart is the thought of mankind" (αἷμα γὰρ ἀνθρώποις περικάρδιόν ἐστι νόημα). In perception, a present object impresses itself on the sense organ, causing the blood to move around the heart, which is thinking.

The ancients identified thought and perception, Aristotle suggests, because they held that both activities discern objects as being a certain way:

> It seems also that thinking (τὸ νοεῖν) and reasoning (τὸ φρονεῖν) are like perceiving something that is (τι εἶναι); for in both of these the psyche discerns something (κρίνει τι) and knows it from among the beings (τῶν ὄντων). Indeed the ancients say that reasoning and perceiving are the same, for example Empedocles says "for it is due to a presence that counsel increases for humans" and elsewhere "whence for them reasoning always becomes differently." And among them, Homer also thought the same thing: "for of such a kind is the intellect." For they all accepted that thinking was embodied (σωματικόν) like perceiving, and that perceiving and reasoning were of like by like. (DA427a19–28)

For Aristotle, Empedocles's claim that thinking is caused by present objects epitomizes the ancient materialists' argument that reasoning and perceiving are the same.

Aristotle himself holds that perception depends on the presence of the object perceived. When one is warmed or reddened, perceiving warmth or redness, something actually warm or red must act as an agent.[2] The misidentification of perception with thinking stems from the fact that thinking, reasoning, and perceiving all seem (δοκεῖ) to be about "something that is" (τι εἶναι). Moreover, thinking and reasoning, like perception, are thought to discern (κρίνει) something among beings (τῶν ὄντων) (DA 427a19–21, cf. AnPo. 99b35). Thus, while it is not propositional, since even animals without language share in this power, it allows an animal to distinguish what is pleasurable from what is painful (DA 432a15–17). Perception discerns one present object from others that constitute the perceptual background. But the senses discern in another way—for example, sight discerns between the visible and the invisible (DA 422a20–21). In this kind of discernment, a visible object is identified as visible. Thus, perception seems to be discerning both in that it picks out a figure from a background, and in that it makes one aware that the object distinguished is perceptible. Thinking and reasoning must have analogous functions insofar as they also put one in touch with reality, with something that is.

Aristotle refutes the identification of perception and thinking with two arguments. First, every animal has perception. Thus, if perception is thinking, every animal thinks. But this cannot be the case since the majority of animals are not rational. Second, if knowledge and perception are alike, then either every appearance is true or contact with the unlike is error (DA 427b2–8). The first option is refuted by counterexample, such as someone's appearing when no one is there (DA 428a12–15, cf. 428b2–4). The second option conflates error, which occurs when a sense organ is in contact with its opposite, with the true knowledge of an opposite. Thus, it seems that the ancient materialist theory cannot account for cognitive error.

Aristotle's conception of the imagination as not being in contact (θίξιν) with its object (DA 427b4) presents an alternative to ancient materialism. Noncontact is a feature that makes the imagination distinct (ἕτερον) from both perception and discursive thinking (DA 427b14–15). In distinction to perception, which is always present (πάρεστι), the imagination does not typically concern present things (DA 428a8–9). Aristotle can therefore distinguish between having a veridical sense perception and having an appearance. When a lion hears the lowing of an ox, it has correctly distinguished a being—its prey—from its surroundings. Merely imagining its food being in some place requires no real connection to the world: something can "appear (φαίνεται) even to those whose eyes are shut" (DA 428a16). As a consequence, one has no reason to trust one's imaginings in the way one trusts one's perceivings: the imagination is "often false" (DA 428a11–12). If perception makes one veridically aware of what is present, imagination makes one falsidically aware of what is absent.

One might, nevertheless, be tempted to think of Aristotelian imagination as a special function of perception. For Aristotle holds that the imaginative part is inseparable from the other parts of the psyche (DA 432a31–b3) and from the body (DA 403a8–11). Moreover, he defines the imagination in terms of actualized perceptual states. It is therefore almost forgivable when Ross calls the imagination "an outgrowth" from the sensitive faculty "on its cognitive side."[3] Yet Aristotle says that imagination and judgment are species of thought (DA 427b27–8, cf. 433a10). Thought here must have general sense, since Aristotle treats intellect in the specific sense of a capacity to grasp universals in the chapters that follow. Nevertheless, since this characterization is placed shortly after Aristotle's refutation of the ancients' identification of perception and thought, it suggests that the imagination is not a function of perception but a distinct power of the thinking psyche.

Comparison to the powers of judgment shows, however, that the imagination is a queer sort of thinking. For instance, belief requires a commitment (πίστις) to something being the case. Since nonrational animals cannot be persuaded by reasons, they are unable to form beliefs (DA 428a20–24, 429a5–6). Similarly, reasoning is discerning (DA 427a20), and the truth of a proposition is determined by a fact (πρᾶγμα) (DA 428b4–9). In contrast, even incomplete animals have imagination (DA 434a4–7).[4] Imagination is distinct from the various species of judgment, since when one is convinced of a proposition, one immediately experiences the corresponding emotion. But an imagined terror is experienced as if "looking at a picture" (DA 427b21–24). Similarly, one can imagine an indefinitely extended line rotating without being convinced of anything and will stop imagining it as other phenomena come to the fore of one's awareness. But when one exercises judgment about that imagined figure, one undertakes a commitment to the proposition that an infinite line cannot describe a complete circle.[5] Barring any countervailing evidence, one is compelled to assent to the same proposition whenever it arises.

Judgments involve real emotional or intellectual convictions, whereas imagined states of affairs are experienced in an "as if" attitude, as merely hypothetical, pictorial, or otherwise unreal. To imagine something is not to make a propositional judgment since "imagination is other than affirmation and negation; for only a combination of *noemata* can be true or false" (DA 432a10–12). Imagination can present things falsely, but strictly speaking it is neither true nor false, since it makes no assertion about how things really are. As opposed to perceptions and beliefs, which relate to beings, an imagined scenario has no evidential status (DA 428a19–21).

The powers of perception and judgment are discriminative, putting one in touch with being and giving one grounds for making true assertions. But imagination is unmoored from reality, producing and presenting one with absent things. As opposed to ancient materialism, Aristotle's psychology can thus explain error as treating something that is merely imagined as if it were present in perception or belief. Yet if this is his response to the ancient materialists, why do Aristotle's modern readers typically maintain that Aristotelian imagination is fundamentally dependent on, or reducible to, perception?

Aristotle's definition of imagination is given as the consequent of a conditional statement: if his division is correct, imagination would be "a motion (κίνησις) that comes about by means of (ὑπὸ) perception when it comes to be active (κατ'ἐνέργειαν)" (DA 428b30–429a4). Despite debate about what imagination is, many hold that this definition requires that every episode of imagination be caused by an episode of activated perception. Frede asserts that imagination "does not have a faculty of its own but is 'parasitic' on sense-perception."[6] Caston characterizes phantasia as an "echoing of the original stimulation of the sense organs."[7] Bolton identifies phantasmata with the afterimages of perception.[8] Although these interpretations differ from one another in many details, they all take imagining to involve a mere reconfiguration of perceptual content.

This common reading has two primary virtues. First, it makes sense of what appears to be an efficient causal attribution in Aristotle's account, by claiming that imaginative episodes are the result of complete perceptions. Second, it makes sense of Aristotle's statement, immediately following the definition, that since sight is perception "most of all," and since there is no seeing without light, the word "phantasia" must derive from the word for light (φώς) (DA 428b30–429a4). This etymological relation of φάντασια to φώς is taken to substantiate the psychological dependence of imagination upon perception.

Yet this interpretation is questionable on both grounds. It is not clear that Aristotle's assertion that imagination occurs in virtue of activated perception requires that perception is the formal, material, or final cause of imagination. Even supposing that perception is the efficient cause of imagination does not entail that the content of imagination derives exclusively from the perception that sets it in motion. Indeed, there is good reason to suppose that a psychological power can have content that is not derived from the power that sets it in motion. Elsewhere in *De Anima*, Aristotle holds that everything is affected and moved by means of (ὑπὸ) a being in activity that acts as the producer of that change (DA 417a17–18). For example, a perceived thing must be something actual, which activates the previously potential perceptual power. But a sphere might be seen as oblong, when illuminated in a certain way. Similarly, the imagination might be set in motion by an actualized perception without that perception producing the content that is imagined. All that is required is that imagination be "triggered" by an active perception. The imagined content itself might be the activity of imagination.

There is also reason to doubt that Aristotle's comparison of imagination to light (φώς) is an etymology. The word "phantasia" is a substantive form of the verb "to appear." And, outside his psychological works, Aristotle uses both forms of the word in an extended or metaphorical sense, as general terms indicative of appearances that might not be real.[9] Aristotle's statement can then be read as analogy rather than an etymology. In the *Ethics*, he gives an example of such reasoning: "As vision is in the body, so intellect is in the psyche, and another in something else" (EN 1096b28–30). Thus, even if he had the mistaken opinion that the word "phantasia" derives from "light," itself a somewhat dubious attribution, the analogy requires just that as light makes possible sight, so perception makes possible imagination. This does not establish that imagination echoes perception but only the weaker point that an animal's having actualized sensory states is necessary for its possessing the power of imagination.

The deleterious effect of the parasitic, "echo" reading of Aristotle's definition of imagination is that it tends to cover up Aristotle's descriptions of the imagination of an image-maker (εἰδωλοποιοῦντες) and of its producing something "before our eyes" (DA 427b19–20). This productive aspect of the imagination depends on its being set in motion by perception but that the imagination itself produces the phantasms that "come to be for us" (DA 428a1–2). In cases of error, Aristotle cannot be placing blame on its perceptual trigger but on the productive and presentational activity of the imagination. One might see one's reflection in a glass, but imagine that there is another person at the end of the hall.

The "echo" interpretation has the tendency to confuse the epistemic function Aristotle ascribes to each part of the psyche. Perception and thinking have epistemic standing due to their discerning among beings of the right kind, particulars and universals. Is imagination also discerning? Aristotle seems to address this question explicitly shortly before making the division between imagination and perception, writing, "is imagination a power or state among those in virtue of which we discern (κρίνομεν), and judge truth or falsity? For those of this kind are perception, belief, science, and understanding" (DA 428a3–5). But in showing that imagination is distinct from these powers, he does not obviously resolve the question. It could be either that the list is incomplete and that imagination is discerning, or that the imagination is a nondiscerning power.

The view that imagination is discerning is widespread. Schofield concludes that it is a "faculty of judgment"[10] and Frede that the imagination is "treated as a critical faculty."[11] The specific discerning functions Frede attributes to the imagination are the synthesis of sense perceptions and the application of thought to the objects of sense perception. According to Frede, the senses "deliver the material that reason works on," material that must be synthesized by the imagination into an image in order to be reasoned about.[12] Thus "from scattered impressions a sufficiently broad orderly picture of the relevant distinctions will gradually emerge that warrants the formation of general concepts necessary for *dianoia*."[13] But how are these purported functions of imagination discerning rather than merely presentational? All that is required is that the imagination present objects as standing in whatever relations the thinking faculty dictates. The requirement on imagination is that it should conform to the critical activity of belief, knowledge, or intellect, not that it discern anything on its own.

The best argument for the view that the imagination is discerning is that of Moss, who holds that evaluative imagination is the "cognition of something as good."[14] In Moss's reading, such "cognition" or discernment is "one of the activities that distinguishes humans and animals from plants" (DA 432a16).[15] Thus, while intellect presents one with what is truly good, imagination presents one with the apparent good. Like intellect, imagination identifies things as good. Unlike intellect, it is apt to be in error.[16] Indeed, in one passage Aristotle states that imagination and perception "hold the same place as intellect insofar as all of them are critical (κριτικὰ γὰρ πάντα)" (MA 700b19–21). If Moss's interpretation is correct, imagination has a critical function in identifying objects as good, thus "rendering them objects of desire" (Moss 2012, 10).

Polansky, in contrast, interprets Aristotle's comment that some animals live by imagination alone and have no rational powers to oppose it (DA 429a4–8) as presupposing a distinction between imagination and the discerning powers.[17] Aristotle can maintain that imagination is connected closely with the powers of judgment and with perception while still distinguishing among them only if the imagination is not discerning.[18] In this view, the role of imagination in presenting one with appearances of objects that are discerned normally in thought and perception can only be understood if one conceives of the imagination as having a dissimilar function.

In Aristotle's descriptions of illusions, the imagination presents contents that can be discerned by perception or by intellect. For example, Aristotle interprets the sun appearing to be the length of one's foot to be a sign that the imagination is something other than the power by which perception discerns (*Insom.* 460b17–19). The cause of this illusion is not that power in virtue of which the controlling sense discerns but "that in virtue of which phantasmata come to be" (γίγνεσθαι) (*Insom.* 460b16–18). The imagination must produce and present one with phantasmata but does not seem to discern any real object. As Benardete puts it, "Touch straightens out the stick we see bent in water; but phantasia has no such diversity within itself to check up on itself."[19]

Adopted from Heraclitus, the sun scenario is a favorite of Aristotle.[20] Facing west at sunset, leg raised to the horizon, it appears that one's foot is the same length as the sun. This is not analogous to a misperception caused by pressing one's eye or chewing miracle berries. Here, one's senses are functioning normally. Nor is the illusion a doxastic error, since it appears even while "one is convinced (πιστεύεται) that the sun is greater in size than the inhabited world" (DA 428b2–4). The illusion that the sun and the foot are the same length can appear even when one has a true belief that opposes this appearance. Yet even when one is in good health and actively looking at the sun, "all the same the sun seems to them to be a foot long" (*Insom.* 458b25–29). As a production of the imagination, the appearance that the sun is the size of the foot can contradict one's veridical perception and true belief.

Aristotelian imagination produces incorrect content in other instances as well. Touching a single object with one's fingers crossed and eyes shut, "one object appears to be two" (τὸ ἓν δύο φαίνεται) (*Insom.* 460b20). In this case, Aristotle explains, while the sense of touch is unable to discern whether the perceiver is in contact with one or two objects, sight is "more authoritative" and can so discern. This is why the illusion dissipates as soon as one opens one's eyes. But Aristotle does not attribute the appearance that there are two objects to touch. Rather, it is the imagination that fills in what touch is unable to discern. The illusion is a case not of misperception but of false appearance.

Imagination continues to function when perception is not actively discerning objects, in cases of emotion, disease, and sleep. Emotions dispose one to see the world in a distorted way. A coward's perceptual disposition is affected by his disposition to experience fear. A lover's expectations are affected by desire. In such cases, a small similarity (μικρᾶς ὁμοιότητος) between a perceived object and the thing one expects to see can lead to the misidentification of the perceived object as that thing. The coward sees the enemy, while the lover sees the object of desire everywhere. The more affected (ἐμπασθέστερος) one is, the less similarity is required for the thing to appear (ἀπ'ἐλάσσονος ὁμιότητος φαίνεται) (*Insom.* 460b3–8). While the central organ of sense normally functions by comparing similarities and differences, emotions dispose one to discern objects inaccurately. The greater the emotional investment, the more biased one's perception.

Aristotle extends this model of affective bias to cases of disease. In fevers one thinks that there appear to be animals around one from their small similarity (μικρᾶς ὁμοιότητος) to lines on the wall, when these are put into patterns (*Insom.* 460b11–13). Like the coward who sees what he fears or the lover what he desires, someone who is feverish is apt to fail in discerning the objects of perception clearly. This affection has degrees, so especially with bad illnesses, the animals in the lines on the wall "move about" (*Insom.* 460b16–17). But while illusions induced by emotions are about something, such as an enemy or lover, in illness the lines on the wall may be simply hallucinatory. Aristotle's description of this hallucination utilizes the productive language from the *De Anima* description of imagination. Just as imagination is said to produce phantasms and put them "before our eyes," here it is said to "put together" or "lay side by side" (συντιθεμένων) the lines on the wall. The product of this activity is the appearance of animals upon the walls, recalling the presentational function of imagination, by which one seems to be looking at a picture.

As in the cases of emotion and illness, sleep makes the capacity for discernment inactive. Thus, "one who is exercising perception is awake"—that is, perceiving "some movement within something else or within oneself" (*Somn.* 454a1–5). But while perceptual activity depends on a central critical power to discern among impressions (DA 424a4–7), in sleep the individual senses are "closed and resting" (*Insom.* 458b7, cf. DA 426b8–24). Sleep is an inactivity in which one has but does not exercise the perceptual power of discernment (DA 412a22–26). This inactive state allows the animal perceptual powers to rest, while the nutritive powers shared with plants become temporarily dominant (EE 1219b23–25, *Somn.* 454b28–455a3, EN 1102a35–b4).

While one rests, a few remaining perceptions can trigger the imagination to produce an image that "has a small similarity" (τὸ μικρὰν ἔχον ὁμοιότητα) to the original perceptual content (*Insom.* 461b10–11). When the sense perceptions are contemporary to the activity of imagination, they produce a hypnagogic experience. One distantly or dimly "hears" the barking of dogs while one is half asleep, but this is fully discerned and becomes clear only upon waking (*Insom.* 462a25). One's hearing is somehow dimmed or mediated, and one does not discern the true volume of the barking and distance of the dog. When the perceiving is temporally prior to the imagining, the perceptual image is retained and "remains" (ἐμμένει) perceptible (*Insom.* 460b2–3), and one has dreams.

Aristotle defines the dream as a "phantasm based on the movement of sense-impressions, when it occurs during sleep *sensu stricto*" (*Insom.* 462a29–31, cf. 462a15–16). Reading this in the context of his affective bias model of illusion, dreams are productions of the imagination, which bear a small similarity to the retained perception that triggers them. Dreams are like shadow-drawings (σκιαγραφία) that present not themselves but something else that is absent (*Met.* 1024b23). They have an intentionality or content but do not appear as dreams. Someone appears to be walking in a dream. But the dream itself is not said to be walking.[21]

The way in which perceptual changes give rise to dreams is discussed in the short treatise *On Prophecy in Sleep* (436a10–15). Aristotle here explains that while many small motions (κινήσεις) go unnoticed during the day, at night such "little things seem large." If these small motions are as sense perceptions, the idea is that subconscious stimuli that are normally crowded out by stronger, conscious stimuli become "enlarged" in the absence of those stronger stimuli. This could explain Aristotle's examples. During sleep, one might suppose that a faint noise in one's ears is thunder, or that a bit of phlegm in one's mouth is sweet honey. In such cases, the perceptual powers present one with content but do not discern it correctly, giving the imagination a dominant role. The faint noise is given by perception but blown out of proportion by the imagination.

The imagination's augmentation of faint perceptions in sleep relates to Aristotle's insistence that dreams can have an intellectual function, when they cause one to think or do something upon waking (*Div.* 463a30). Aristotle is of the view that "just as in waking life we think discursively" (διανοούμεθά) about things other than perceptions, so "also in dreams we have something in mind (ἐννοοῦμεν) aside from phantasmata" (*Insom.* 458b18–19). What one has in mind is not identical to one's dream. The dream serves as a vehicle for thinking, a way of rehearsing what one has perceived and thought about in waking life.

Aristotle is consistent in claiming that both perceptual and intellectual forms are represented in imagination (DA 431a16–17, 431b2, 432a8–15, 433b29; DM 702a19; *Mem.* 449b30–450a5; *Insom.* 458b18–19). But this status as a thinking power has been downplayed by readers who assume a "parasitic" understanding of the imagination, particularly in the case of dreams. Ross, for example, describes perception as a motion of the soul through the body that "sets up a repercussion both in the body and in the soul." This repercussion is only potential and unconscious "until recollection takes place," at which point it becomes actualized and conscious. Thus, "at some later time, owing for instance to the suppression of sensation in sleep, the movement becomes actual," where the movement in question is "an image similar to but less lively than the sensation, and less trustworthy as a guide to objective fact is formed and attended to; and this is the act of the imagination."[22] Ross assumes a Humean interpretation, in which the contents of imagination are reproductions of sensory stimuli, and thus takes dreams to consist entirely of perceptual motions retained in the body of the animal.

Bolton has a similar reading, carried out more systematically. There is a similarity between the afterimages that persist after the external cause of a perception has departed (AnPo. 99b36–100a4) and Aristotle's characterization of the phantasma as "a retention of an impression present in the activity of perception" (*Insom.* 461b21–22). Moreover, Aristotle describes perceptual afterimages as affections or movements of the blood. Bolton concludes that phantasmata must also be movements in the blood about the body.[23] Thus, "this phantasmic state is what constitutes for Aristotle the material cause of the perception in question, the internal state that arises in the organs themselves during perception."[24] Phantasmata for Bolton are copies of perceptual afterimages that circulate in the blood and are discerned by the central perceptual organ of the heart.

It is unlikely that this was Aristotle's considered view. Aristotle describes dreaming as coinciding with a flood of blood to the heart, which disables perceptual discernment. Ross himself observed that "the critical faculty is in abeyance owing to the pressure of the blood on the heart, the central organ of perception."[25] How could a disabled critical power grasp the phantasmata? Bolton is surely right that Aristotle's account of dreaming is a part of his natural science. But the movement of blood around the heart is a material account of the weakening of discernment, not of its discernment of phantasmata. Sleep consists in having but not using the discerning power of perception. When the stronger stimuli of perception are removed, the imagination takes over. Thus, dreams are caused by a lack of perceptual discernment.

Aristotle's physiological theory presupposes the imagination as the power that produces something that has a "small similarity" to what one perceives. Aristotle thus gives imagination an active role when the perceptual power is inactive. Hence it is due to imagination that "something appears also when neither potentiality nor actuality is attributed to (ὑπάρχοντος) perception, such as in sleep" (DA 428a6–8). Imagination here is not taken to reproduce actualized perceptions but to function as an independent productive and presentational power of the psyche. Upon waking up, to be sure, one can sometimes recognize that an image one experienced while asleep has as its substance a movement in the sense organ (*Insom.* 462a8–12). In this sense, the imagination "follows upon" perception (συμβαίνουσαι ἀπὸ τῶν αἰσθημάτων) (*Insom.* 461a19, cf. DA 429a1–2,427b14–16; *Mem.* 451b21–22).

Could this "following upon" support the stronger thesis that the content of phantasmata are completely determined by one's antecedent perceptual experience? Moss points out that imaginings are similar to the activity of perception, since Aristotle describes imagination as a motion occurring "only in perceivers and for whom there is perception, since too it is possible for motion to occur as the result of the activity of the perceptions, and this is similar (ταύτην ὁμοίαν) as in perception" (DA 428b10–14, cf. DA 429a5–6). But similarity does not amount to identity. Had Aristotle intended to identify actualized perceptions with the states of imagination, or to treat phantasmata as copies of them, one should expect the phantasmata, rather than the power of phantasia, to be the antecedent to this description. Moss also claims that Aristotle's comparison of a memory caused by a perceptual experience to the imprint of a signet ring supports the view that the contents of perception and imagination are the same.[26] This argument depends on identifying the "imprint" (τύπον) of sense with the phantasm (*Mem.* 450a25–32). But why should an "imprint" entail that there is a physical similarity between a perceptual object and its representation in phantasia? A semblance in some respect imitates, but in other respects differs from, what it represents. Similarly, an episode of the imagination might be quite different from the perception that triggers it.

In an extreme case of the "parasitic" interpretation, Johansen concludes that the imagination is not a distinctive power at all. Johansen's argument depends on three theses. First, imagination is a kind of motion that comes about only as a result of the activity of perception. Second, imagination has the same objects as perception. Third, imagination is not an efficient cause.[27] For Johansen, the imagination is neither a cause like perception nor does it have its own states. Since everything found in the psyche is either an affection, state, or power (EN 1105b20–22), it follows that imagination is an affection of the psyche. In this view, the imagination is wholly inactive and parasitic on perception, and thus cannot present intelligible forms. This is supported by a passage in which phantasmata seem to be treated as belonging to the perceptual part of the psyche. "Necessarily, we know magnitude and motion in the same way we know time, and the phantasma is an affection of common sense, so that it is clear that knowledge of these is due to the primary perceptive part. But memory, even the memory of intelligible things, is not without a phantasma. So, memory would belong to discursive thinking accidentally, but essentially to the primary perceptive part" (*Mem.* 450a9–15). According to Johansen, this establishes that remembering consists of relating a present phantasm to the past. The imagination "serves its purpose in memory precisely by retaining what has been perceived as it was perceived." And because *phantasia* deals with the same content of perception, "we can infer that *phantasia* belongs to the perceptual capacity, rather than some other capacity." Thus, for Johansen, the imagination is a mere remnant of perception, which "retains the information conveyed, the sensible form, when the object has moved on."[28]

Magnitude, motion, and time are common sensibles. Knowledge of these originates in the primary perceptive part. But it does not follow that a phantasma is produced by perception. Rather, in light of what I have argued, it is more likely that Aristotle considers the phantasmata to be a modification of these previously given objects. Consequently, the dependence of memory on phantasmata of time shows not that imagination is the product of perception but that perception alone is insufficient for memory. Thus, when one recollects a discursive concept rather than a sensory quality, the imagination is required to place that concept in a time-relation. The function of the imagination in this passage is to produce an appearance in which something accidental to common sense, a general concept, or "memory" of intelligible things is grasped in accordance with common sense. As argued in the following chapters, Aristotle makes an analogous argument for magnitudes, which are given perceptually but may also be reasoned about geometrically when the imagination serves to divorce them from their perceptual context.

The object of memory is not the phantasma itself but the absent object that the phantasma represents (*Mem.* 450b11–16). Far from making the production of phantasmata a function of memory, Aristotle supposes that the imagination must produce a phantasm in which a previously encountered perceptual content is presented as a distinctive content. Memory requires that one depict in an image something that was previously discerned, thereby making the thing an object of contemplation (*Mem.* 450a20–27). The function of imagination in this account is not to give one the same perceptual experience one had before but to produce and present a depiction of the absent thing in such a way that one thinks of it as past.

This depictive role of imagination is also in Aristotle's treatment of the recollection of atemporal, intelligible objects, which are only "accidentally" encountered in perception.[29] Intelligible things are only accidentally objects of memory, but the recollection of intelligibles also requires memory (*Mem.* 451b5–6). Thus, a purely theoretical concept counts as an object of memory insofar as it is depicted in a phantasma (*Mem.* 451a1–3). For an atemporal object of intellect to be put in a time-relation, and hence to be related to perceptual experience at all, the imagination must produce a phantasma from which the intelligible content can be grasped. If memory is a state or affection of perception or judgment when time has elapsed (*Mem.* 449b26–28), the imagination is necessary to retrieve and depict the content recorded in memory.

I have been arguing that Aristotle's use of the imagination in his *Parva Naturalia*, while primarily concerned with the psychological phenomena associated with perceptions, dreaming, and memory, makes room for intellectual and epistemic functions of the imagination. While this is perhaps unsurprising, given that Aristotle holds that the objects of intellect—the noemata—must be thought "in" phantasmata, it has been glossed over by commentators who tend to view Aristotle's work as defending a naturalism that is more related to British Empiricism than it is to ancient concerns about the interplay of perception and intellect in producing knowledge. But how does Aristotle understand this epistemic role of imagination? The following chapter attempts to answer this question by examining Aristotle's cognitive account of geometry, the canonical example of ancient science.

CHAPTER 5

ABSTRACTING

Natural philosophy is about things that "are separate but not motionless," while mathematics concerns those things that are "motionless but not separate" and "equally in material" (*Met.* 1026a13–16). Aristotle holds that the circle is combined with matter (*Met.* 1035a34–b1) and that some matter "is sensible, some intelligible" (*Met.* 1037a4–5). So perhaps mathematical objects occupy a "third realm" intermediate between things in sensible, moving matter and the universals grasped by intellect. Yet outside the *Metaphysics*, intelligible matter is not mentioned and seems to be of little import to Aristotle's theories of geometrical inference and cognition. Though it is possible that intelligible matter was a matter of debate within the Academy, there is insufficient evidence to determine whether it was part of Aristotle's ontology of mathematical entities. This difference between perceptible and mathematical magnitudes is thus taken to constitute a serious puzzle in Aristotle's ontology. How can mathematics not concern perceptible things but also not concern other, separate things besides perceptible things, either (*Met.* 1078a5, cf. 1077b12)?

Mueller holds that mathematical objects must be real for Aristotle, since they are precise geometrical properties derived from abstraction instantiated in intelligible matter.[1] Lear claims that mathematics separates geometrical properties from the perceptible matter and thus counts as a fictional treatment of real properties that physical objects actually have.[2] Netz argues that when Aristotle describes mathematics as being about sensible magnitudes "not qua sensible but qua magnitude" (*Met.* 1077b22), he employs "qua" (ᾗ) adverbially rather than adjectivally. Thus, mathematics makes no peculiar ontological commitments but operates within a "perspective" in which its inferences are taken to be true. This supports a conditional realism on which, if reality matches the perspective, the mathematics developed within that perspective may be taken to be true of the world.[3] Corkum holds that mathematical objects can exist apart from sensible substances but that they have this ontological status only in virtue of standing in relation to sensible substance.[4]

These views are focused largely on the impasses Aristotle establishes at the opening of *Metaphysics M* (1077a1), which are taken to constitute a debate between a naive Platonism that posits mathematical objects as wholly independent of perceptible matter and a naive materialism according to which the mathematical things are in matter (*Met.* 1076a32–35). Aristotle argues, against the materialist, that were mathematical things divisible material bodies, then points would be in divisible bodies. But points are indivisible. Therefore, if a body were divided, then the indivisible would be divided (*Met.* 1076b5–12). Against the Platonist, he objects that the multiplication of entities is unhelpful to science. If one must posit a separate object domain for each mathematical science, then there would be a "second heaven" studied by astronomy, apart from the perceptible one, and the object domains of optics and harmonics would consist of sights and sounds that would require new senses of sight and sound, different from those we actually possess (*Met.* 1077a1–8).

Aristotle's arguments are, however, mainly concerned to show that neither ontological program is helpful to mathematics. And his ultimate reason for rejecting the naive views betrays an altogether non-ontological purpose: "on the whole what results is contrary both to the truth and to what it is customary to assume, if one posits that mathematicals are by nature separable things" (*Met.* 1077a14–16). This is not an argument from first principles but a point about convention and practice. It is not customary to assume that mathematical objects are separable. Aristotle's claim is not about the truth of an ontological scheme but about the nature of science. He aims to explain the actual procedure of the geometer.

Aristotle's understanding of geometrical practice is informed by his logical distinction between the proper and accidental subject matter of a science. For example, a biologist might suppose that every sexed animal either is male or female. But these affections are not separable from being an animal of that kind and are therefore are "accidental." To say that an animal is male or female does not contribute to an account of what it is to be that animal. Analogously, geometry is concerned with lines and planes only insofar as they are lines and planes, and not insofar as they are perceptible. From the point of view of geometry, perceptibility is accidental. And it is in virtue of this that geometry is simpler and more exact than natural science (*Met.* 1078a2–11).

In characterizing a science that ignores the accidental features of its objects as being more exact than science that concerns accidental features, Aristotle makes both a logical and a cognitive point. Logically speaking, there is a correspondence between the exactness of a science and its ignorance of accidents. The more features are ignored as accidental, the greater the generality of its statements will be. Cognitively speaking, what is more familiar to us is "nearer to perception" while what is prior by nature is "further away" (AnPo. 72a1–5), so that more exact science will have less to do with perception. Thus, in characterizing geometry as an exact science, Aristotle suggests that it is cognitively "simpler" and closer to intellect than the empirical sciences are.

The ontological debate concerning mathematical objects is therefore irrelevant to geometry. For just as being bronze is accidental to a thing's being a sphere, material qualities are accidental to the science of geometry. This can be seen in Aristotle's definition of the accident in the *Metaphysics*: "Accidental means that which is asserted of something and is true to say of it, but it is not so either by necessity or for the most part, such as if someone who dug a hole for a plant discovered treasure. For this discovery of treasure is accidental to digging a hole for it. For it is neither by necessity that one comes from the other or with the other, nor for the most part that one who is planting something discovers a treasure" (*Met.* 1025a14–19).

To grasp mathematical objects within perceptible magnitudes is like finding a treasure when one is digging a hole for a plant. Just as one finds something that is not related to planting in digging the hole, one strikes upon geometrical properties accidentally when one perceives a magnitude.

Yet there is a difference between the treasures discovered in planting and those discovered in geometry. Whereas digging neither always nor for the most part results in discovering treasure, geometrical search can have the force of necessity. Hence "accidental" has a secondary sense when it is "attributed to each thing in virtue of itself but not in its substance, for example, that the triangle contains two right angles. These may be everlasting, but the others are not" (*Met.* 1025a30–33). What it is to be a triangle is to be a three-sided, closed planar figure. Yet containing two right angles is an everlasting property of every triangle. The necessary accident does not constitute the substance or essence of the triangle, since "having two right angles" is not what it is to be a triangle. But because it is necessary, even though geometrical assertions are made of a substance, the universal demonstrative content of the science of geometry can be thought without any substance (AnPo. 79a7–9). Unlike the planter, the geometer can set out to find treasures without ever thinking about the substances in which they inhere.

Recasting the procedure of Menaechmus as a logical doctrine rather than a constructive program, Aristotle understands geometry as a demonstrative science that deals with its subject matter though the use of elements. An element is "that out of which something is composed, as the first constituent not divisible in kind into a different kind" (*Met.* 1014a25). The elements of geometry are like those of other demonstrative sciences in that they can be used recursively to prove different propositions, such that "the first demonstrations are also present in the rest of the demonstrations, and these are called elements of the demonstration" (*Met.* 1014b1). This syllogistic structure serves as a kind of formal "memory" that gives a form to the progressive procedure of demonstration. Geometrical properties are simple, everlasting, and precise, and thus logically prior to material particulars. But to make inferences about them, one must represent them in some substance to which they are accidental, typically in figures drawn in sand or on paper.

Diagrams play an important role in ancient geometry, as is evident in Aristotle's description of a sophistical objection to geometry. A geometer, having just drawn a line of several feet in the sand, points to the line and asserts that "this line is one foot." The geometer then constructs a figure on the line—for example, an equilateral triangle. From the fact that the base of the equilateral is one foot, the geometer infers that each of the other sides of the triangle are likewise a foot long. A sophist objects that because the original line drawn is not really a foot long, the premise is false. One should not accept the geometer's conclusion, either. Aristotle replies that the geometer is considering the figure as separated from its accidental properties. Though the line is in fact not really a foot long, the geometer is not referring to the line as perceptible. The falsehood is not part of the premises (*Met.* 1078a20–21). Geometrical inference concerns only the intelligible attributes of the figure, excluding its perceptible attributes from the deduction. The objection that the line is not really a foot fails because it does not respect the convention of the geometers. Aristotle holds, then, that geometrical properties should be given a verbal treatment that differs from ordinary discourse, and which serves to separate them from the other properties of a figure in which they are instantiated.

Yet for Aristotle, geometrical procedure is not solely a matter of language. In postulating the line as separate, the geometer engages in a sort of "vision" of geometrical diagrams that warrants inferences that are independent of the sensible material in which the diagrams are inscribed. Indeed, Unguru had gone so far as to maintain that any purely algebraic reconstruction of ancient geometrical reasoning, because it rests on a form-content distinction that is foreign to Greek mathematics, necessarily fails to explain actual Greek mathematical practice.[5] According to Unguru, the custom of "translating" geometrical propositions consisting of texts and diagrams into symbolic equations is intellectually irresponsible since it does away with the very historical phenomena it proposes to illuminate. Ancient geometrical practice consists of an interplay between the diagram and the text of a proposition. The diagram serves as an embodied graphic representation of spatial thinking, while the textual component of the proof serves a hortatory, descriptive, and rhetorical function insofar as it specifies how to obtain "finished, polished, wholesome diagrams."[6] While algebraization obscures both the spatiality of the diagram and the prescriptive function of the accompanying text, ancient geometrical practice always involves both the immediately spatial component of the diagram and a normatively binding textual component that specifies just how the geometer ought to conduct veridical spatial reasoning.[7]

Unguru's suggestion that diagrams' constitutive role in geometrical inference illuminates Aristotle's otherwise mysterious concern with the relation between mathematical cognition and the artefact of the diagram. In the *Sophistical Refutations*, Aristotle points out that one often fails to recognize a known fact when it is laid out (μετατιθέμενον) differently and argues that one's speed in recognition depends on training. Just as failure to demonstrate a proposition occurs (συμβαίνει) with geometrical diagrams and we "analyze a figure but cannot synthesize it," so also in dialectical refutations "we know how the argument is strung together (συνεῖραι), but we are at a loss about how to divide it" (SE 175a22–31). Aristotle here assumes that diagrams are necessary for geometry and draws an analogy between the layout of diagrams and the construction of dialectical argument. Insofar as diagrams and arguments are similar, they should involve similar cognitive activity. Similarly, in the *Ethics*, Aristotle argues that "the deliberator is like one who inquires (ζητεῖν) and analyzes (ἀναλύειν) in a manner, as if by diagram (τρόπον ὥσπερ διάγραμμα)" (EN 1112b20–24). The term "analysis" and its verbal and adjectival cognates originate in a mathematical context and originally mean "working back from the result aimed at" to the principles required for the result.[8] In both Aristotelian passages, analysis means a decomposition of a constructed figure into its elements.

The main extant source on ancient geometrical analysis, the introduction to book 7 of Pappus's *Collection*, portrays analysis as essential to geometrical technique. In his account, the study of analysis and its counterpart, synthesis, gives one a power of discovery by which geometrical problems can be resolved. "Analysis is the path from what one is seeking, as if it were established, by way of its consequences, to something that is established by synthesis. That is to say, in analysis we assume what is sought as it has been achieved, and look for the thing from which it follows, and again what comes before that, until by regressing in this way we come upon some one of the things that are already known, or that occupy the rank of a first principle."[9] Pappus then distinguishes between two sorts of analysis: in theorematic analysis, one seeks the truth, whereas in problematic analysis, one seeks what is demanded in the statement of the construction problem. Thus, Pappus goes on to make a statement that is in tension with his initial description.

> In the case of the theorematic kind, we assume what is sought as a fact and true, then, advancing through its consequences, as if they are true facts according to the hypothesis, to something established, if this thing that has been established is a truth, then that which was sought will also be true, and its proof the reverse of the analysis; but if we should meet with something established to be false, then the thing that was sought too will be false. In the case of the problematic kind, we assume the proposition as something we know, then, proceeding through its consequences, as if true, to something established, if the established thing is possible and obtainable, which is what mathematicians call "given," the required thing will also be possible, and again the proof will be the reverse of the analysis but should we meet with something established to be impossible, then the problem too will be impossible.[10]

In the first passage, Pappus suggests that in analysis what is assumed hypothetically is a consequence, and what is sought are the premises that would make it true. In this case analysis is a "reduction" of a problem, whereas the following synthesis is a deduction of the desired result. In the second passage, he suggests that what is sought are the consequences that would follow from a hypothetically assumed premise. In this case, analysis is itself deductive.

There is a long-standing debate concerning how to resolve this tension in Pappus.[11] One attractive view is that Pappus countenanced both problem reduction and the deductive search for consequences under the broad heading of "analysis."[12] But without opining on this contentious issue, one can observe that this tension or confusion between analysis as a search for preconditions and analysis as a search for consequences is already present in Aristotle. In the passages in *Sophistical Refutations* and the *Ethics* just mentioned, Aristotle seems to take analysis to refer to the preconditions of the construction. Indeed, as argued later on, Aristotle's comparison of the deliberator to the geometer rests on the presupposition that both search for the means of bringing about their desired end. Yet in the *Posterior Analytics* (78a6–14), Aristotle describes analysis as a kind of deduction. "If it were impossible to demonstrate truth from falsehoods, it would be easy to perform the analysis [of arguments], for then they would convert necessarily. Let A be something that is the case; and if A is the case, then these things are the case (things which I know to be the case—call them B). From the latter, then, I shall prove that the former is the case. In mathematics conversion is more common because mathematicians assume nothing incidental (and in this too they differ from those who argue dialectically) but only definitions."[13] In contrast to his characterization of analysis as a search for preconditions, Aristotle here treats it as the search for consequences. The tension that modern readers have detected in Pappus is thus already present in Aristotle.

This tension can be partially alleviated, in the case of Aristotle, by observing that his comments in both the *Sophistical Refutations* and *Posterior Analytics* are comparisons between dialectical and geometrical analysis. In the dialectical case, propositions do not normally convert, so one cannot safely assume that knowing the consequences will be sufficient for demonstrating the premise. In contrast, geometrical analysis is performed in abstraction from the incidental features of the figure. The geometer, unlike the dialectician, assumes only definitions that convert. It is thus typically possible to reason from B back to A, without the risk of a fallacy entering into the geometer's reasoning. Though I shall have more to say shortly about how this abstraction is performed, it is at least clear that Aristotle here assumes the same doctrine that he used to refute the Sophists' objections to geometry. The geometer may make a number of false statements about his diagrams, but this falsity does not enter into the premises of his demonstrations. Aristotle therefore treats analysis as a procedure of geometrical discovery, rather than deduction proper. Cognitively, this process is a doxastic "expansion," insofar as it works "backward" from a completed figure to its elements, producing an ordered set of definitions or constructions. At the next stage of synthesis, these elements are reversed to obtain the completed construction, and hence to demonstrate the proposition.

Proclus's claim that Plato originated the method of analysis and taught it to Leodamas (*InEucl.* 211) is widely regarded as false, since Hippocrates of Chios probably used the method as early as 430 BCE.[14] Nevertheless, Aristotle here follows a trend in the Academy to take geometry as a paradigmatic model for the sciences.[15] Used in problems of construction, the method of analysis consists in assuming the figure one is seeking to be constructed and then working to find the constructible elements of the thing sought.[16] But regardless of whether analysis univocally names a single method in Aristotle or in the later geometrical tradition, it is at least clear that Aristotle took diagrams to be constitutive of geometrical inquiry.

The act of seeing in a diagram serves as a primary tool of Greek geometrical reasoning. For instance, Manders argues that in Euclid's *reductio* proofs, diagrams are used to represent constructions that violate conditions given in the discursive text. Thus, the use of diagrams in *reductio* proofs "does not presuppose that everything asserted in the discursive text be 'true' of the diagram" but in fact depends on a dissonance between the text and diagram.[17] This incompatibility makes evident the impossibility of constructing the desired figure. In Greek geometry, diagrams did not *illustrate* the text but were positively employed in making geometrical inferences.

What distinctive features do the text and diagrams possess? According to Manders, exact attributes, which in modern mathematics are expressed by algebraic equations, "are those which, for at least some continuous variation of the diagram, obtain only in isolated cases; paradigmatically straightness of lines or equality of angles."[18] In Greek geometry these are typically given by equalities or proportionalities in the text of the proposition and are destroyed with the variation of the figure. For example, if one side of a triangle were extended, the angle opposite the extended line would increase, while the adjacent angles would be reduced proportionally. Co-exact attributes "express the topology of the diagram" and pick out a region, segment, or point from the whole. In ancient geometry, attributions of co-exact attributes are licensed directly by the diagram and survive some class of topological variations. Expressed by strict inequalities, they are locally invariant under perturbation of the diagram.[19] For example, if one side of a triangle were extended, the number of sides of the triangle would remain the same. What Manders calls "attributes," Aristotle would call "accidents."

Aristotle's main philosophical concern about geometrical inference

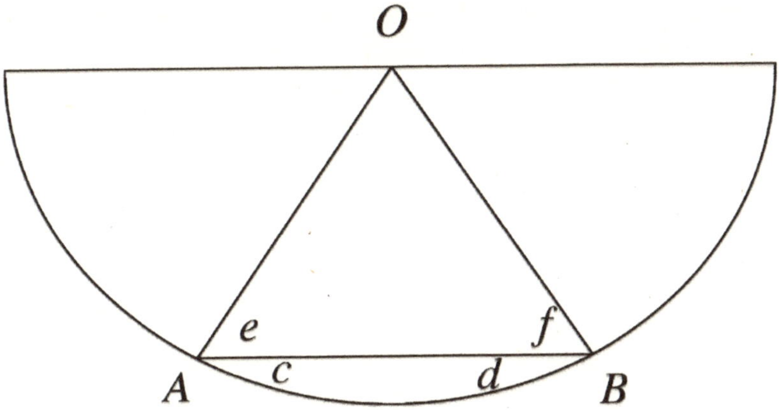

FIGURE 3

is to establish its generality, which requires the use of universal propositions. Thus, in *Prior Analytics*, he argues that if one's aim is to prove that pleasure in music is good, the proposition that some pleasure is good cannot be a premise. Nor could the proposition that pleasure in music is good be a premise, since this would be to assume the conclusion. The only way to infer that something is good, is to begin with a universal premise that is not identical to the conclusion. Thus, "if one should claim as a premise that pleasure is good without adding 'all,' no syllogism will be possible." Expanding on why a universal premise is required, Aristotle turns to the geometrical proposition that the angles at the base of an isosceles triangle are equal.

> Suppose the lines A and B have been drawn to the center. If then one should assume that the angle AC is equal to the angle BD, without claiming generally that angles of semicircles are equal; and again if one should assume that the angle C is equal to the angle D, without the additional assumption that every angle of a segment is equal to every other angle of the same segment; and further if one should assume that when equal angles are taken from whole angles, which are themselves equal, the remainders E and F are equal, he will beg the thing to be proved, unless he also states that when equals are taken from equals the remainders are equal. It is clear then that in every syllogism there must be a universal premise, and that a universal statement is proved only when the premises are proved, while a particular statement is proved both from two universal premises and from one only. (*AnPr.* I.24, Jenkinson, trans.)

Although Aristotle's point is clearly about the kinds of universal inferences made in the discursive portion of geometrical propositions, the theorem as stated assumes one already has a certain geometrical figure in view. Aristotle and his auditors would likely have used figure 3.[20]

Aristotle's universal statements about equalities of angles and lines depend on looking at a diagram that provides the co-exact properties of the geometrical figure. The figure contains information necessary for the proof: the lines A and B are connected or "drawn" at the origin of the semicircle. The angles formed by lines A and B are taken to contain C and D, respectively. Thus, when C and D are subtracted from A and B, of which they are parts, the remainders are E and F. Connection, containment, and remainder are topological, co-exact attributes of the diagram that must be seen for the inference to be complete.

Aristotle's point is that taken alone these co-exact attributes of containment and connection do not justify the inference. Assuming knowledge of these diagrammatic relations, he claims that demonstration requires universal propositions that are not contained in the diagram. The demonstration is (1) the angles A and B are equal, and (2) the angles C and D are equal common parts of the angles B and D, respectively. Thus, when (3) the equal common parts of each angle, C and D, are removed, (4) the remainders E and F are equal to each other. Where is the universal premise? Aristotle holds that the last step of this deduction depends on the common notion, "When equals are subtracted from equals, the remainders are equal." In other words, Aristotle is arguing that mathematical statements must be made in the demonstrative language of his syllogistic. Scientific propositions are only established through universal premises. Yet diagrams provide a special tool for geometry, insofar as they present topological information not contained in universal discursive statements.

The universal logical form of geometrical demonstrations, combined with the necessity of using diagrams to obtain such demonstrations, gives rise to a philosophical quandary. A diagram always presents some particular triangle or some particular circle, not the universal triangle or universal circle. How could it warrant universal geometrical inferences?

Aristotle addresses this question in *Posterior Analytics* with reference to the proposition that the angles of every triangle are equal to two right angles. How can a universal proof be achieved, given that one always perceives either an equilateral, isosceles, or scalene?

> Even if you prove of each triangle either by one or by different demonstrations, that each has two right angles—separately of the equilateral and the scalene and the isosceles—you do not yet know of the triangle if it has two right angles, except in the sophistic fashion. Nor do you know

> it of a triangle universally, not even if there is no triangle apart from these. For you do not yet know it of the triangle as triangle (except in respect of number; but not of every one in respect to kind, even if there is none which you do not know). So when do you not know universally, and when do you know *simpliciter*? Well, certainly you would know *simpliciter* if it were the same thing to be a triangle and to be an equilateral (either for each or for all). But if it is not the same but different and it belongs as triangle, you do not know. Does it belong as triangle or as isosceles? And when does it belong in virtue of this as primitive? And of what does the demonstration hold universally? Clearly whenever, after abstraction, it belongs primitively. For example, when two right angles belong to the bronze triangle, but also when being bronze and being isosceles have been abstracted. But not when figure or limit have been. For they are not the first. But what is the first? If triangle, it is in virtue of this that it also belongs to the others, and it is of this that the demonstration holds universally (An Po. I.5., 74a17–74b4).

Aristotle holds that it is in virtue of "abstraction" (ἀφαίρεσις) that one can infer not just that some bronze triangle has two right angles but that this is true of the triangle simpliciter. Abstracting from accidental features such as being bronze or being equilateral is supposed to render universal knowledge of the triangle. Aristotle describes abstraction as the "casting off of everything perceptible," an operation that leaves only a continuous quantity. The only relations remaining among the abstracted contents are commensurabilities and incommensurabilities (*Met.* 1061a28–b3). It appears that abstraction is the operation that makes universal inferences in geometry possible.

Commentators have pointed out that the cognate verb for abstraction (ἀφαιρεῖν) literally means "to eliminate" or "to take away" something, so that abstraction is best understood as "subtraction." In Plato's works, the verbal form usually arises in a pair with the dialectical method of "adding on."[21] Cleary argues that these terms gained a technical meaning initially in a dialectical context, probably in the Academy, but even when used technically, they have primarily a logical rather than psychological or epistemological significance.[22] Reviewing the main instances of the term in *Posterior Analytics* and *Metaphysics*, Mueller concludes that "abstracting involves eliminating something particular from consideration. This is not a matter of collecting particulars and somehow arriving at a general idea."[23] This logical procedure of elimination has no ontological consequence, since "it is clear that the thing that results from taking something away is not prior, nor is the thing that results from adding something derivative, since it is by addition to the white that the human being is articulated" (*Met.* 1077b9–11). The proof that the triangle simpliciter has two right angles is explanatorily prior to the proof that the scalene has two right angles, since the scalene has this property in virtue of its being a triangle. That does not entail that the triangle simpliciter is prior in being to the scalene.

How does abstraction license universal mathematical inferences? The psychological interpretation takes abstraction to be a mental act by which one removes an attribute from an inner representation. Given a mental image of a galloping black Arabian horse, one eliminates its galloping, its blackness, and its Arabian breed. What is left is the general concept of "horse." Classically, such psychological abstraction entitles the nominalist, who only accepts the existence of material particulars, to use general concepts. This is Locke's understanding of abstraction.

> For, since all things that exist are only particulars, how come we by general terms; or where find we those general natures they are supposed to stand for? Words become general by being made the signs of general ideas: and ideas become general, by separating from them the circumstances of time and place, and any other ideas that may determine them to this or that particular existence. By this way of abstraction they are made capable of representing more individuals than one; each of which having in it a conformity to that abstract idea, is (as we call it) of that sort.[24]

Maintaining his ontologically nominalist convictions, Locke argues that one is entitled to general type terms like "horse" and "triangle" as mental abstractions of particulars. Thus, he takes abstraction to be a psychological operation that removes a particular representation from its original, perceptual circumstances. Though there are no types in the world, abstraction produces a general mental representation to which the type term refers.

Though some commentators in antiquity seem to attribute such a view to Aristotle, modern readers typically resist this interpretation.[25] For if the triangle in general is just a mental representation, abstracted from this bronze triangle, how could one's abstract concept determine the truth about every triangle? The only possible answer, on the Lockean interpretation of abstraction, is that all universal inferences depend on geometers believing them to be true. This amounts to a gross psychologism in which thinking that a truth about the scalene is true of the triangle simpliciter makes it so.

Avoiding such psychologism about mathematics, more recent interpretations have argued that abstraction is primarily a semantic notion. In Lear's view, for instance, Aristotelian abstraction "amounts to no more than the separation of one predicate that belongs to an object and the postulation of an object that satisfies that predicate alone." Because it is logical, it escapes the psychologistic consequences of the Lockean interpretation. Consequently, though the separation of the predicate may occur in thought, "this is no more damning than admitting that one carries out a geometrical proof or arithmetical calculation 'in thought.' One is carrying out a determinate procedure and there is no irremediably subjective element."[26] In this view, abstraction is fundamentally a linguistic operation.

It is not immediately evident how taking away the bronze from a bronze triangle is a linguistic act. But Lear holds that when one treats a magnitude in a perceptible diagram "as if separable," one employs a "*qua*-operator." This operator serves as a "predicate filter" that removes predicates irrelevant to the mathematical demonstration. Thus, "by applying the predicate filter to an object instantiating the relevant geometrical property, we will filter out all predicates which concern the material composition of the object. Thus the geometer is able to study material objects—this is indeed all he studies—but does not study them as perceptible or as material."[27] In this view, the Aristotelian separation of mathematical from perceptible magnitudes amounts to "taking away" those predicates that relate specifically to the perceptible nature of the magnitude. Consequently, the "simplicity of abstract knowledge" is "achieved through the postulation of separated objects." Knowing the property of the triangle simpliciter results from postulating a triangle without the predicates "scalene," "equilateral," and "isosceles." Hence abstraction not only separates a magnitude from perception but also entitles more general inferences, since the fewer predicates are included in a description, the more universal and simple the statement will be.[28] In Lear's view, abstraction is a semantic elimination procedure independent of psychology.

According to this semantic interpretation of abstraction, every mathematical object must be somewhere instantiated in a physical object. But Aristotle holds that geometrical figures are infinite in number (*Sens.* 442b22). It would follow that there are an infinite number of physical objects in the cosmos, which instantiate an infinite variety of geometrical properties. Moreover, supposing that one of these bodies were destroyed, how could the corresponding geometrical construction be possible? Worse, the semantic interpretation leads to an unattractive explanation of the scenario in which a geometer utters of the inch line on paper that "this is a foot." If "this" refers to the diagram, then the proposition is false. If it does not, it follows that while pointing and making statements about one material particular—the diagram—the geometer

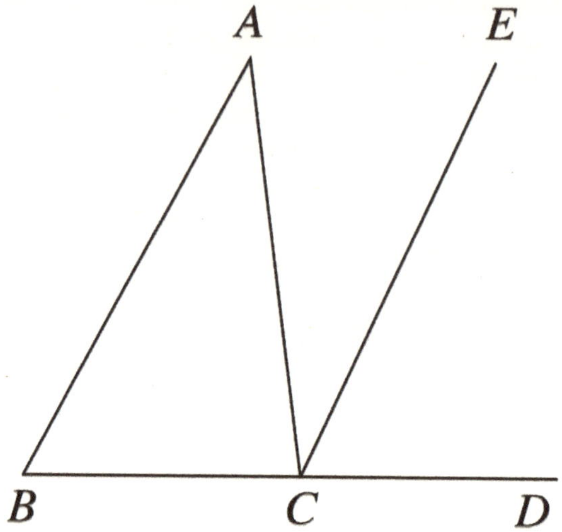

FIGURE 4

is actually referring to some other, distal material particular. But the fundamental problem with the view that abstraction is semantic is that the discursive demonstration is only part of a geometrical inference, the other part being achieved through the construction of a diagram. This diagrammatic component of geometry is simply ignored on the linguistic account of abstraction. Yet knowledge that the triangle in general has two right angles depends on the geometer's being able to read off necessary accidents from a nonlinguistic, perceptually given diagram. Universal results are, like a treasure buried in soil, discovered in diagrams by the diligent work of the geometers.

Aristotle illustrates this conception of the treasure of universal results buried in particular figures with an example: "Why does a triangle have two right angles? Because the angles about one point are equal to two right angles. So if the base were drawn beyond the side, this would be clear immediately to the one who looks at it" (*Met.* 1051a24–26).[29] In figure 4, beginning with any triangle, the crucial move is "drawing beyond the side"—that is, drawing lines from C to D and to E. This step allows for the crucial comparison of interior and exterior angles of a triangle.

To see how the diagram proves the universal proposition, one must be able to view the triangle as standing in for any triangle whatsoever. The triangle printed on the page has a particular material and shape. In abstracting from it, one treats it as the universal triangle. This becomes evident when one considers that the triangle can be topologically manipulated however one wishes, without changing the result. The only requirement for the proof to hold is that CE must be parallel to AB.[30]

This example shows that abstraction can be identified neither with the psychological production of a general idea from mental representations of particulars, nor with the linguistic operation of predicate elimination. For neither of those procedures relates to the topological manipulation of diagrams necessary for the general result. Rather, it is the imagined perturbation of co-exact features that allows one to treat *this* triangle as standing for the triangle universally. Though one draws a scalene, isosceles, or equilateral, it is the deformation of the diagram in the imagination that reveals the generality of the proof. The process of deformation displays how the particular triangle drawn is irrelevant to the proof that the triangle as triangle has two right angles. This diagrammatic procedure of elimination of irrelevant alternatives reveals that the general proof is potentially "in" the particular diagram. When the co-exact attributes—the figure's "accidents"—are perturbed, the exact attributes of the figure may be grasped in their actuality.

Abstraction, in effect, presents this particular triangle as the universal triangle, insofar as it reveals that diagrammatic topology is accidental to the proposition. Abstraction is thus not a mental process, since what makes the proof true is not a mental representation. Like the semantic conception, this interpretation contains no irremediably subjective features. But nor is abstraction a purely linguistic procedure, since it involves a kind of "production" that alters and deforms the diagram. Yet the perceptually given diagram is never concretely moved. The abstracting step of the proof only treats the diagram as if it were being deformed. Psychologically speaking, then, the spatial reasoning required for geometry is neither completely perceptual nor completely discursive. Rather, the production and presentation required to grasp geometrical demonstrations must be the functions of imagination.

CHAPTER 6

GEOMETRICAL THINKING

Plato's views on mathematics are difficult to reconstruct with accuracy. Yet one thesis that was at least entertained within his Academy was that mathematical objects are independent of the material particular in which they are represented. For example, in the *Rival Lovers* (132a–b), a Platonic author describes two youths in a grammar school who are arguing (ἐρίζειν) about an astronomical theory of Anaxagoras or Oenopides, over circles they have drawn (γράφειν), while imitating (ἐμιμοῦντο) motions with their hands. But, Socrates maintains, while diagrams are necessary for the practice of pure and applied mathematics, they are not sufficient for the justification of general statements. Like gestures representing the motion of the sun, the diagram of the circle is a mere imitation of the circle itself. Likewise, in the *Euthydemus* (290b–c) Socrates argues that although they might not recognize it, geometers, calculators, and astronomers do not simply produce diagrams (οὐ γὰρ ποιοῦσι τὰ διαγράμματα) but discover beings (τὰ ὄντα ἀνευρίσκουσιν). For these beings to be correctly categorized, the practitioners of these sciences must hand over their results to dialecticians for division and collection, just as fishermen hand over their catch to cooks. Philosophy must cook up geometrical discoveries, giving them a correct ontological interpretation.

There remains a nagging concern. How can geometers discover beings that are imperceptible by the senses from a perceptually given diagram? A geometer's description of looking at a diagram cannot give such knowledge, since his perception is limited to that particular diagram. This is why, in the *Republic* (527a1–b1), it is said that the geometers fail to describe their own practices correctly. Although they use terms like "adding," "squaring," and "applying" to describe their procedures, they speak absurdly. Diagrammatic procedures are a necessary cognitive aid for the discovery of mathematical beings, but these beings exist independently and can be grasped as such only by the dialectician. Plato's philosophy of geometry thus characterizes diagrams as imitations of independent mathematical objects, akin to the hand-waving of the youths in the grammar school. Geometry, calculation, and astronomy are therefore merely heuristic enterprises, whose true objects are grasped only by the philosopher.

Aristotle rejects this dualism between mathematical objects and their graphical representations while maintaining that mathematical results are in fact universal. Part of this rejection is motivated by metaphysical worries concerning the multiplication of entities.[1] But he is also concerned that his epistemic claims about geometry accord with the practices of the geometers, which he understood to depend fundamentally on drawing diagrams. As argued in chapter 5, Aristotle employs the concept of abstraction to explain why the geometers' procedures are logically warranted. In a few passages, however, he offers metaphysical and cognitive reasons for his conception of geometry as a demonstrative science.

Aristotle's conception of geometry as a science of universals rests on his view in the *Metaphysics* that geometrical constructions are discovered in the activity of division (εὑρίσκεται δὲ καὶ τὰ διαγράμματα ἐνεργείᾳ: διαιροῦντες γὰρ εὑρίσκουσιν). A diagram "contains" a number of truths in potentiality, which are not immediately apparent (φανερά) but become apparent in the division (*Met.* 1051a21–24). Division brings about geometrical discovery insofar as grasping relations among the salient parts of the figure presents the intelligible content of the diagram to the intellect. Assuming the proposition that the angles of a triangle are equal to two right angles, Aristotle illustrates this with an example.

> Why is the angle in a semicircle universally a right angle? If three lines are equal, two forming the base, and one set upright from the middle of the base, the proposition is clear from beholding the figure and knowing the property. For it is apparent that potential beings are discovered

> by being drawn out into activity. But this is because thinking is activity. So potentiality proceeds from actuality, and it is on account of the making [of the diagram] that we come to know [the proposition]. For in generation, the activity according to number is posterior (διὰ τί ἐν ἡμικυκλίῳ ὀρθὴ καθόλου; διότι ἐὰν ἴσαι τρεῖς, ἥ τε βάσις δύο καὶ ἡ ἐκ μέσου ἐπισταθεῖσα ὀρθή, ἰδόντι δῆλον τῷ ἐκεῖνο εἰδότι. ὥστε φανερὸν ὅτι τὰ δυνάμει ὄντα εἰς ἐνέργειαν ἀγόμενα εὑρίσκεται: αἴτιον δὲ ὅτι ἡ νόησις ἐνέργεια: ὥστ᾽ ἐξ ἐνεργείας ἡ δύναμις, καὶ διὰ τοῦτο ποιοῦντες γιγνώσκουσιν ὕστερον γὰρ γενέσει ἡ ἐνέργεια ἡ κατ᾽ ἀριθμόν). (*Met.* 1051a26–33)

It is hard to say whether one should translate *energeia* in this passage as "activity" or as "actuality." Activity picks up on the notion that drawing diagrams out by hand is how one learns to demonstrate a geometrical proposition. In this reading, to engage in geometrical cognition is to determine in generation numerically, so that the diagram one actively produces reveals the number of cases a problem may take, depending on how many ways the definitions can be drawn out on paper or in the imagination. Translating energeia as "actuality" suggests that the passage concerns discovery, in which an unknown becomes known. The cause of this is that the actuality of intellect is noesis. The power or potential to φ, in this view, depends on the actuality of φ-ing. If no one could φ under any circumstances, then there would be no actuality, and thus no power to enact φ-ing. In this reading, actuality is prior to potentiality in generation because there are only so many ways a formula or definition can be actually satisfied.

These considerations of language relate to more concrete interpretive difficulties encountered in this passage. What figure did Aristotle have in mind—or on display in the lecture hall—when he composed this paragraph? What is it that is clear from observing the diagram? What is in the diagram, or presented via the diagram, that Aristotle considers potential, and what actual? On the side of cognition, what is potential and what is actual? What is being generated, and in what sense is one to understand this generation?

Makin discusses three possible geometrical interpretations of this passage.[2] I favor Heath's interpretation, which takes Aristotle's formulation to differ from the more general proof in proposition III.31 of Euclid's *Elements*.[3] Assuming that in a circle the angles in the same segment are equal to one another,[4] Aristotle would be reasoning as follows.

> The line OC bisects AB, so $\angle AOC = R$ (i.e., is a right angle). Since they are radii of the same circle, AO = OC. Thus, $\triangle AOC$ is isosceles.

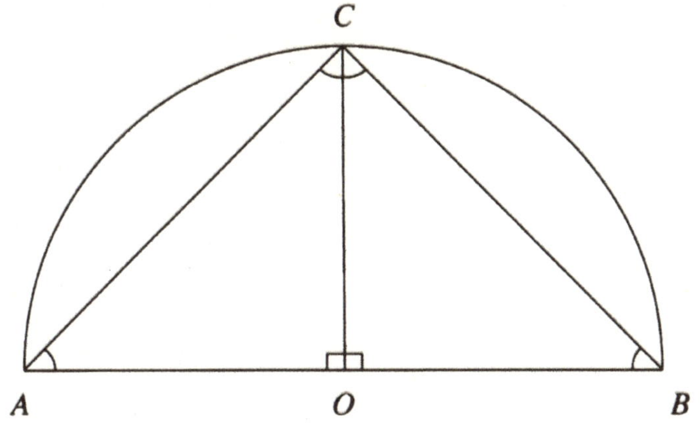

FIGURE 5

Angles at the base of an isosceles triangle are equal, so ∠OAC = ∠ACO. But since the interior angles of a triangle are equivalent to two right angles and ∠AOC = R, then ∠OAC + ∠ACO = R.

Similarly, OC bisects AB, so ∠BOC = R. Since they are radii of the same circle, OB and OC are equal because they are both radii of the circle. Thus, △BOC is isosceles.

The angles at the base of an isosceles triangle are equal. Therefore, ∠OBC = ∠BCO. The interior angles of a triangle are equivalent to 2R. Because ∠COB = R, and ∠OBC = ∠BCO, then ∠OBC + ∠BCO = R.

Therefore, ∠ACO + ∠BCO = R, but ∠ACO+∠BCO is equivalent to the angle in the semicircle ACB. This proves that the angle in the semicircle is a right angle.

The diagram corresponding to this reconstruction are given in figure 5. Figure 6 shows why Aristotle can utilize the property that angles in the same segment of a circle are equal to one another: when the point C is "moved" along the semicircle, the angle ∠ACB is always right.

This interpretation rests on taking ἐκεῖνο to refer to the property established in the previous proposition—namely, that the angles of a triangle are equal to two right angles.[5] This interpretation is appealing both geometrically and cognitively. This reading links the two propositions under discussion via a middle term—the property of being equal to two right angles—that is present in both propositions. Cognitively, this takes Aristotle to be attributing to the diagram the function, familiar to anyone who has worked through some of Euclid's proposition, of making one aware of the geometrical property relevant to proving the proposition in question. The diagram presents a geometrical property, some-

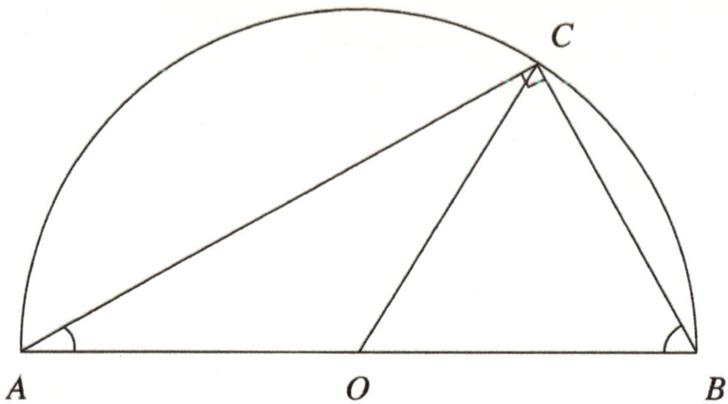

FIGURE 6

thing that is actually universal and intellectual, as a potentiality clothed in a particular figure.

Aristotle is more concerned with the proposition's deductive form in the *Posterior Analytics* (94a28–36). There, he argues that since each of the four "causes" or forms of explanation must be proved through a middle term, scientific explanations can only be exhibited through a demonstrative syllogism. Barnes holds that Aristotle uses this proposition to exemplify the form of explanation that corresponds to material cause, which is marked by the phrase "if something holds it is necessary for this to hold." Linking this mode of explanation to Aristotle's concept of intelligible matter,[6] Barnes claims Aristotle means to indicate that "the angle in the semicircle is constructible out of the halves of two right angles, which are thus its 'matter.'"[7]

While I am unsure of the identification of the "matter" of the premises with intelligible matter, the identification of this logical mode of explanation as material makes sense of Aristotle's discussion of potentiality and activity in geometry in the *Metaphysics* account of the proposition. The inference is from property of the triangle to that of the semicircle and depends on a shared middle term. Suppose one takes this property to be the matter of the triangle. Then before it has been inscribed in the semicircle, the property is only potentially in the semicircle. Inscribing the triangle draws that material or potential "into actuality," in that it then becomes apparent that the semicircle in fact has the same property. In this sense, the diagram may be thought of as the realization of equivalent properties in the triangle and semicircle.

Mathematical truths are independent of generation but are grasped through a generative process. In the geometrical case under consideration, the proposition that the angle internal to a semicircle is universally a right angle is discovered through the construction of a diagram, a "beholding" of the diagram, and an activity of intellect that eventuates from these processes. While the universal is logically prior, the produced material particular is prior with respect to perception, and in the temporal order of discovery (*Met.* 1018b29–34). Makin's claims that Aristotle is committed to a "quasi-perceptual intuition" being required for geometrical cognition.[8] But this is unclear. Aristotle surely conceives of geometrical cognition as requiring one to perceive a diagram and to understand its rational bearing on the proposition in question. But this is a matter of making the correct inference from a given diagram, not of enjoying an intuitive, non-inferential grasp of the proposition. By viewing the diagram in the proper attitude, the geometer's cognition is aroused by his perception, and he understands in actuality the geometrical truths the diagram contains potentially.

Mathematical beings qua intelligible are potential objects of intellect. They have the capacity to be understood but are merely potential beings while they are not understood. The activity of intelligible understanding (ἡ νόησις) is the actuation of their being. But this actuation depends on leading or drawing off the intelligible aspect from the diagram and presenting it before the intellect. As argued in the last chapter, the intelligible aspects of a diagram are the necessary accidents salient to what is sought in the geometrical demonstration. To be presented to the intellect, these accidents must be divided from the nonsalient features of the particular material diagram. Aristotle attributes this division, in which the salient features are read off the diagram, to the imagination. The plausibility of this attribution depends on recognizing that Aristotelian imagination is capable of both production and presentation.

Aristotle's *Metaphysics*, insofar as it treats actualities, places more requirement than his other texts upon his theory of the psyche. Aristotle describes the diagram as having a potential to be actuated in the geometer's grasp of the universal. But neither perception nor intellect is sufficient to bring about the required change. Moreover, Aristotle insists that use of the language of potentiality in geometry is only metaphorical (*Met.* 1019b33–34), a thesis perhaps based on his view that mathematical objects are not subject to change (*Met.* 1026a13–16). But if geometrical cognition requires no actual change in the material of the diagram nor in the mathematical being itself, how can the perception of a material particular give rise to an act of understanding a universal?

If presenting part of the diagram as salient to a universal proof is the work neither of perception nor of intellect, what is required is a third cognitive power, the imagination. In his *De Memoria*, Aristotle gives his most extended description of the role of imagination in geometrical cognition.

> An account has already been given about imagination in "on the psyche" that there is no thinking without a phantasm. Indeed the same affection corresponds to thinking and to drawing (συμβαίνει γὰρ τὸ αὐτὸ πάθος ἐν τῷ νοεῖν ὅπερ καὶ ἐν τῷ διαγράφειν). For in the latter case, though the determinate quantity (τὸ ποσὸν ὡρισμένον) is not pregiven in [the triangle], all the same we draw it as determinate (γράφομεν ὡρισμένον) with regard to its quantity. And it is the same in thinking as well, that even when one is not thinking something as a quantity, a quantity is nevertheless placed before the eyes, even though one does not think the quantity. (*Mem.* 449b30–450a5)

According to this characterization, the imagination connects the particular grasped by perception with the universal content of the inference. The perceptual manifold is the field from which the imagination divides the salient part. But the imagined object is not itself a perceptible, material particular. Rather, the phantasm produced by the imagination is an affection of the perceptible particular that makes the drawing susceptible to universal inference. The intellectual process of thinking the universal, in which there is no specified determinate quantity, thus emerges from imagining a modification of the diagram, an activity that draws out its salient features, giving them a determinate quantity. In order to read off geometrical relations from a diagram, the imagination must produce a depictive phantasm in which the intelligible can be grasped by the intellect.

Reasoning by diagrams according to Aristotle requires a threefold conception of cognition. Perception is necessary for seeing the diagram and discerning among its parts. Intellect is necessary for the activity of knowing the universal content of the geometrical proposition. Finally, imagination serves as an intermediate power that places the determinate quantity of the diagram to the understanding, making its thinking possible.

Plato's dualistic view of diagrammatic discovery and philosophical knowing separates the perceptual activity of the geometer from the universal reasoning of the dialectician. To be sure, the geometer will have discovered universal results through his investigations, but these remain merely hypothetical until they are radically clarified by the dialectician. Geometry thus has a purely heuristic value prior to its completion by philosophy. In contrast, Aristotle's threefold distinction among the cognitive processes underlying geometrical discovery aims to explains straightforwardly how geometers come to know the propositions of geometry.

Perception is the power that establishes a relation among magnitudes (DA 418a18–20, 424a25–28) such as surfaces and lines. But when treated theoretically, magnitudes are separated from all perceptual characteristics. Mathematical objects are thought to exist separately from their perceptible matter, and thus when they are being thought, they do not exist separately from the power that thinks them (DA 431b15–17). Thus, physical bodies do not lack mathematical properties altogether, since they contain surfaces, volumes, lines, and points, the things examined (περὶ ὧν σκοπεῖ) by the mathematician. But as opposed to the natural philosopher, the mathematician does not treat the objects as limits of some particular body. Rather he separates a magnitude from its accidents, "for it is separable from motion in thought (χωριστὰ γὰρ τῇ νοήσει κινήσεώς ἐστι), and there is no difference, nor does falsehood come to be when things are separated (οὐδὲ γίγνεται ψεῦδος χωριζόντων)" (*Phys* 192b8–193a8). Perceptible things are potentially objects of thought but must be separated from the perceptual manifold in which they are given in order to be actualized mathematically.

Taking perceptible objects as starting points in this way does not suggest that the truths of mathematics depended on perceptible things. For "the objects of thought are in the forms that are perceived, since it seems that there is no thing (πρᾶγμα) separated from the perceptible magnitude." Thus, things considered in abstraction (ἐν ἀφαιρέσει) are states and affections separated from the objects in which they are perceived (DA 432a3–6). The right angle instantiated in a wooden frame is not something different than the angle determined by the meeting of two sides of definite length. But the perceptual grasp of the right angle is altogether different from the theoretical study of the right angle in geometry. Thus, "a carpenter and geometer study the right angle differently; for one does it as much as it is necessary for his work, but the other studies what it is or how it is, for he would see what is true" (EN 1098a29–32). The carpenter, the representative of the productive arts, has inexact, practical knowledge of the right angle. Similarly, "it is for the physician to know that circular wounds heal more slowly, and for the geometer to know the reason why" (AnPo. 79a15). The arts of carpentry and medicine depend on a body of knowledge or observed facts that allow them to achieve their desired ends. Geometry, however, gives the ultimate reasons for these facts.

As a theoretical science, geometry concerns intellectual forms. These primary thoughts are present neither in perception nor in the imagination but are objects of intellect (DA 432a12–14). Nevertheless, thinking is always accompanied by an image produced by the imagination (DA 431b2). Thus, in order to come to universal judgments about geometrical objects in mathematical science, one must present such objects to oneself in the imagination but employ the intellect in making inferences about those objects (DA 432a8–9). In abstraction, the imagination is allowed to vary the given perceptual image, producing a continuous range of topological deformations and ending after a number of distinct topological "motions." In imaginative variation, the phantasma serve as a content about which the universal judgment can be made, playing the role of a perceptual object without actually being instantiated in matter (DA 432a9–10).

Aristotle envisages geometrical cognition as consisting of three distinct stages. First, in an act of paradigmatically visual perception, one receives the perceptible form of the common sensible of magnitude. This perceptual act is shared by both the geometer and the carpenter. Second, one re-presents the content of the original act, depicting the magnitude in the imagination. This depiction, now separated from the original perceptual magnitude in which it occurred, makes possible an intellectual inspection of the properties of the magnitude "in abstracto." Images of geometrical objects are here inspected as standing in certain relations or as exhibiting certain properties. Finally, one makes inferences about the geometrical object presented in the imagination. This is the activity or actuality of intellect in grasping the proposition, which Aristotle calls "*noesis*."

None of these processes commits Aristotle to attributing mathematical truths to the imagination. On the contrary he holds that imagination is "often false" (DA 428a11–12) and that it is the discursive part of the psyche that is responsible for making the assertions and denials in the proof (DA 431a14–17). Thinking, rather than imagination, discerns the truth and makes the relevant deductions. Imagination plays an auxiliary, productive, and presentational role in geometrical cognition.

If what I have argued is correct, the universality of a geometrical judgment is ensured by the elimination of irrelevant alternatives. Reaching such a judgment depends on varying the dimensions of the diagram, a process that gives a range of definite quantities to that which lacks a definite quantity. This topological variation could not occur within the drawn diagram, which any perturbation would in fact destroy, but rather must occur in the imagination of the geometer. Thus, imagination is essential to mathematics not because mathematical objects are fictions—Aristotle treated all the discoveries of geometry as unqualified truths—but because it allows the geometer to interact with a diagram fruitfully. The imagination allows for the construction, reinterpretation, and the removal of irrelevant alternatives through deformation of the diagram. Hence, whatever their ontological status may be, the cognitive status of geometrical objects must be intermediate between perceptual and intellectual forms, a rank mirrored by the placement of mathematics between the sciences of physics and first philosophy. Geometry treats of objects that are simpler and more universal than things that are grasped by perception but which are more familiar and nearer to us than things grasped by the intellect.

CHAPTER 7

WISHING

Had Aristotle restricted his use of the imagination to the theoretical domains of psychological explanation and geometrical inference, the power might have remained in the preserve of pure philosophy. But Aristotle observed an analogy between theory and practice that made the application of imagination in ethics irresistible. In the theoretical domain, the imagination is needed to put unchanging objects in motion, enabling one to make general judgments from the observation of particulars. Similarly, in the practical domain, actions are always particular, concrete events. How could reason be brought to bear on one's choices, which contribute to an end that is not directly observable but might only ever exist as a mere possibility? Seeing that the imagination could explain how one can present oneself with hypothetical and counterfactual states of affairs in the theoretical sciences, Aristotle takes imagining possibilities, comparing them, and connecting them to one's actual state of affairs as the hallmark of practical rationality. Imagination thus turns out, for Aristotle, to be what makes it possible to conceive of the end one ought to pursue in life.

Imagination is not necessary for action, though. For Aristotle considers desire to be what moves every animal to act. All that is required to form desires is that an animal possess the power of perception. "If an animal has the perceptive part it also has the desiderative part. For desire is appetite, temper, or wish, and all animals have at least one sense: touch. And perception is attributed to this thing; for it there is immediately both pleasure and pain and the pleasant and painful, and for these there is also appetite, for this is the desire for the pleasant" (DA 414b1–6). Every animal can desire something present, insofar as it can form an appetite for something in its immediate perceptual environment. In human beings, however, appetite, temper, and wish can be distinguished (EE 1223a27–29). Wish (ἡ βούλησις) is in the calculating part of the psyche (ἔν τε τῷ λογιστικῷ), while appetite and temper are in the irrational part (ἐν τῷ ἀλόγῳ) (DA 432b5–6). Thus, one can be moved by desire contrary to calculation or in accordance with calculation, which Aristotle also calls being moved "according to wish" (κατὰ βούλησιν) (DA 433a22–26).

How could a form of desire be distinctively rational? One might want to pursue a course of action upon which one has deliberated. If that course of action is the right one to bring about one's aim, one's choice (προαίρεσις) would be a rational desire. But Aristotle distinguishes choice from wish according to their objects: "A wish is of the end rather than the means, while choice is of the means relative to the end" (EN 1111b25–30). Choice and wish thus name two different forms of practical rationality. Taking the course of action one has determined to be the most effective way of achieving one's end is rational, in the sense of choice. But a wish is not instrumental to one's ends. Rather, it is required for having an end beyond the satisfaction of one's immediate bodily and emotional needs. Wish is thus rational in the sense that it gives one a reason to act, over a period of time, to achieve some goal.

There is an ambiguity, however, in the relationship between one's wish and one's end. Value realists claim that wish is for the good. Value relativists claim that wish is for the apparent good (EN 1113a15–16). The dilemma that emerges from the comparison of these positions is a neat puzzle. But it raises a serious question. Given that people pursue many discordant aims, how can one understand one's own aim as good, and thus as giving one a reason to act?

The realist, following Socrates, claims that failure to pursue the good cannot be the result of a wish, since one wishes only for what is really good. Thus, "for those who say that the thing wished for is the good, it follows that there is not a thing wished for when one wishes for what is chosen incorrectly. For if it were a thing wished for, it would also be good, but as stated, it happens to be bad" (EN 1113a17–19). The realist takes wish to aim at the actual good and argues that no one knowingly does what is bad for himself.[1] For the realist, failure to achieve a good end must be attributed to a false belief. So anything one took oneself to want, but which turned out to be bad, was not really one's wish. It merely appears that people wish for a multiplicity of opposed ends.

The relativist, following Protagoras, claims that one may wish for whatever seems good.[2] Thus, "for those who say that what is wished for is the apparent good, nothing is by nature wished for, but only what strikes (τὸ δοκοῦν) each person as good. But since things appear (φαίνεται) different to different people, these may happen to be contrary" (EN 1113a20–22). The relativist holds that whatever appears or strikes one as good is actually good. Since people pursue a multiplicity of opposed ends, there is no single good that "is by nature wished for." Rather, each person has his own private good, which he authoritatively defines for himself.

Aristotle objects to both views. The realist is stuck with the unattractive view that people wish for the good itself rather than what they take themselves to wish for. For instance, though one might think one wishes for a glass of wine, one really and inadvertently wishes for something else—namely, the good. This runs afoul of the intuitive notion that in consciously wanting something, one ipso facto knows what one wants. The relativist has other issues. Suppose some end appears good to one person but bad to another. Since both appearances are true, it follows that it is both good and bad. Thus, the relativist must hold that contraries are true of the same thing at the same time and in the same respect.

Aristotle started with the idea that one can have desires that are rational, in the sense that they give one a reason and not just an inclination to act in a certain way. This gives rise to the dilemma about the apparent good. Thus, it seems that the discussion of what it means to pursue the good necessarily runs into the old dichotomy between appearance and reality, the very domain the imagination was designed to explain in Aristotle's theoretical philosophy.

Aristotle seems to deny that reasons cause actions when he writes that "nothing that is not desiring or avoiding is moved, except by force" (DA 432b15–17). Unlike inanimate objects and plants, living animals are put into motion by desire, an affection of the animal psyche (EN 1106a8–13, 1105b22, EE 1220b11–20). In Aristotle's Greek, as in modern English, desire can have an occurrent or a dispositional sense. One might dispositionally desire good health but occurrently desire a cigar. In describing motivating desire as an affection, I think, Aristotle means desire in an occurrent sense. When an animal is directed to a nearby object in an evaluative way, that object acts as the mover of the animal (DA 433a23–28, MA 700b24–29).[3]

Inanimate objects and animals both have dispositional states in virtue of which they display law-like regularities. But the tendency of glass to shatter or of mercury to travel up a tube during a thunderstorm are not affections of their forms like desire. Desire is distinctive in that, when activated, it produces a formal relation that can obtain between a living animal and the thing it wants.[4]

What distinguishes cases in which an animal is moved by force, such as by a strong wind, from those in which it is moved by the object of desire, such as by a source of food? Force acts on the body, while desire acts on the psyche, the form of the animal. But since both can lead to the same outcome—the wind might move an animal to a source of food—an account of self-movement must be based on something other than the achievement of a goal. Aristotle thus distinguishes three parts of the animal that contribute to its motion. The good that can be done (τὸ πρακτὸν ἀγαθόν) produces the movement of the animal. The desire, by which the motion comes about, in potential is called a "mover" but remains unmoved; when actualized it becomes a motion. Finally, there is what is set in motion, the animal itself (DA 433b13–18). Animal movement thus consists in a formal producer, the material body of the animal, and their combination, the moving animal.

Due to this special relation between form and matter, the simple notion that desire causes animal motion has to be qualified. A thing's appearing as desirable is sufficient to produce an occurrent desire, so that when an animal is "activated towards something by perception or imagination or intellect," it immediately moves (MA 701a29–30). As an affection rather than a power or state of the psyche, desire must be activated by a something that can be obtained, a "practicable good" such as a pleasant object (EN 1106a8–13, 1105b22). Yet because the animal body is complex, no one thing is always pleasant, but different things are desirable at different times (EN 1154b20–24). An animal's self-motion, insofar as it is caused by desire, is determined by what appears good or pleasant to that animal.

Aristotle identifies the formal cause of movement as the object of desire, which has the formal property of seeming good to the animal. Irwin has argued that calling something an "apparent good" does not commit Aristotle to the view that the object appears to the animal *as* good, so that Aristotle might simply mean to indicate that something appears to an animal and happens to be good for an animal.[5] However,

Moss points out that the description of bad aims appearing to be good to an animal are explanatory of why an animal should seek them.[6] Thus, denying that the apparent good refers to something that appears good to an animal loses the very explanatory power of the concept of apparent good.

In the case of wish, the rational form of desire, this dependence of desire on the something's seeming to be good is more explicit. Wishing requires that I suppose (οἴομαι) something to be the case. Supposition can indicate an incomplete doxastic commitment so—for example, older men have less conviction in what they and others say and consequently say not that they are sure but only that they suppose something to be true (*Rhet.* 1389b17). Supposing can designate an unwarranted assumption, so that when a necessary middle term is lacking in a syllogism, one merely supposes that one knows the conclusion without really knowing it (AnPo. 75a15–16). And supposing can express expectations or intentions—for example, when what happened was not what he supposed would happen (EN 1053b14–15). But in describing wish, Aristotle uses supposition to refer to a judgment to which one is committed, though one lacks apodictic proof of its truth.

> "Now wish is desire for the good—for no one wishes but when he supposes (οἰηθῇ) the thing is good; temper and appetite are irrational desires." (*Rhet.* 1369a2–5)
>
> "Everything that one does [when one is] wishing to do that thing, is voluntary; and that which is wished is voluntary; but no one wishes what he supposes (οἴεται) to be bad." (EE 1223b5–9)
>
> "Those wishing (βουλόμενοι) for a life together (συζῆν) with friends, do the same things on account of which things they suppose (οἴοντα) life together consists." (EN 1172a6–8)

Wishing coincides with supposing something to be the case. But supposition is not veridical. According to the statements considered here, acting voluntarily to bring about an end depends on having the thought that this end is good. If that end, after one had voluntarily brought it about, turned out to be bad, one's supposition was wrong.

Recalling that Aristotle placed wish in the calculating or rational part of the psyche, I suggest that the supposition involved in wishing is the conceptual content of the wish. If this is correct, then to wish to bring about this end is to hold that this end is good in such a way that if all barriers to bringing about this end were removed, one would be committed to acting immediately to bring about this end. Overt statements of suppositions thus count as articulations of the conceptual content of one's wishes.

This is strikingly different from Humean and Kantian theories of practical reasoning. Hume's view is that passions are "original facts and realities, compleat in themselves," that motivate action.[7] Passions tell us what we want; reason how to get it. In distinction, Kant held that "the practical rule is always the product of reason." Here, it is reason that does the commanding, and will either obeys or disobeys. Thus, "if reason completely determined the will, the action would inevitably take place according to this rule."[8] Hume and Kant agree that of will or reason, one must be the master and the other a slave. They only disagree about how to organize the internal economy of the moral subject.

In contrast, in Aristotle's conception, wish is both wholly rational and wholly desiderative. Internal conflict is possible when appetite or temper usurp wish's rightful place in the kingdom of the psyche. But within wish, there is no tension between reason and desire, because they are not conceived as different capacities with respect to action. Aristotle articulates this thought in at least one passage, in which he writes that "what assertion and denial are to discursive thought, these are pursuit and flight to desire" (EN 1139a21–22). Pursuit and flight are a desiderative form of assertion and denial, a sort of instruction one gives oneself, which is implicit in the action that the desire initiates. In the cases of appetite and temper, this instruction cannot be conceived as propositional, so they are merely analogous to the kinds of assertion and denial performed in discursive thinking. But wish is rational in that it has a conceptual content that can be discursively articulated in the statement of one's supposition. Far from cobbling together distinct notions of reason and desire, Aristotle's view recognizes that one's reasons for acting just are the conceptual content of one's desires.

The realist denies that appearances could influence what one wishes for, while the relativist identifies appearance with reality. Both come to the same remarkable conclusion that wish is necessarily for the good. Aristotle dissolves this dilemma by pointing out that wishes can be for what is bad, just when one has mistaken suppositions. But what could lead one to make such a mistaken proposition? In analogy with cases of illusion, illness, and dreaming, Aristotle suggests that the imagination can lead one to make faulty practical judgments. Thus, what an animal should pursue or flee from "is defined in phantasmata," since even in the absence of perception, the animal may be "set in motion when the phantasmata are before it" (DA 431b3–5). Indeed, Aristotle may be committed to the stronger claim that all desire requires a presentation of the object as an apparent good, and thus as desirable (MA 700b24–9), and that animal motions always come about "with imaginings" (μετὰ φαντασίας) (DA 432b15–17). Combined with the idea that desire aims at an apparent good (MA 700b28–29), these views suggest that an end appears as good due to the exercise of the imagination and thus that the apparent good is the object of wish.

The imagination's freedom from determination by concrete particulars serves an explanatory role in his theory of geometrical cognition. Consequently, one might think that when Aristotle's proposition that we imagine something "whenever we wish" (DA 427b18–21), he means that one can imagine whatever one wishes to imagine. If one wishes to imagine one's father, one would immediately produce and present oneself with an image of one's father.[9] But how should one know what to wish for, unless one already had an image of that thing in mind? Here a regress of images looms: one would require an antecedent image of one's father, and so on ad infinitum.

One can avoid attributing a regress of images to Aristotle by reading the statement conversely, as indicating that imagination is constitutive of wish. Indeed, Aristotle describes instances in which one imagines what one does not wish to imagine—for example, in the cases of illusions and dreams. The sun's seeming to be the length of one's foot is not something one wants to imagine but a phantasma that arises independently of one's desires. The claim that one imagines whenever one wishes more plausibly means that imagining a state of affairs is necessary for wishing for that state to obtain.

This reading is supported by Aristotle's view that imagining is a necessary ingredient of intellectual processes like grasping a geometrical proof, since the psyche never thinks without a phantasm (DA 431a16–17). If imagination is employed in all thinking, it seems that it must also be employed in wishing, the form of desire in the thinking part of the psyche. Indeed, desire, whether rational or irrational, causes an animal either to pursue or avoid something: the objects of wish, temper, and appetite are the fine, expedient, or pleasant; their contraries, the shameful, the harmful, or the painful, are the objects of avoidance (EN 1104b30–32). Moreover, while the process of deliberation and its product, choice, are about the means, wish, temper, and appetite are "of an end" (EN 1111b26–27, cf. 1112b11–12, 1113a15, 1113b3–4, EE1226b2–5, 1227a3–5). Deliberation and choice are evaluatively constrained both by an agent's orientation to their goals and modally constrained by what is in their power to do. Thus, "if someone appeared to choose the impossible, he would seem to be an idiot; but wish is also for impossible things, such as immortality" (EN 1111b20–23). Though the imagining involved in deliberating and choosing is modally constrained by the end one pursues and by what one takes to be possible, there are no such constraints put on the imagination when one merely wishes for some state of affairs to come about.

Wish is not constrained by one's self-interest, either. The wish of the statesman, for instance, "is to make the citizens good and obedient under the law" (EN 1102a9–10), an outcome that is primarily good for the citizens rather than for himself. For the function of the statesman is to produce friendship, which is necessary to make good citizens (EE 1234b22–30, 1103b3–6). The friendships of others, the object of the statesman's wish, is an essentially altruistic aim. Nor are wishes constrained by one's being in a position to bring them about, since "a wish can also be of things which might be done not by the man who wishes them, like the wish that a certain actor or athlete be the victor" (EN 1111b23–24). Any imaginable state of affairs might be an object of wish.

Yet wish is a necessarily evaluative activity: "No one wishes what he does *not think to be good*" (EN 1136b7–8). Consequently, the statement that we imagine whenever we wish must be refined to indicate the evaluative orientation of the imaginative act of wishing. Aristotle's statement thus means that whenever one wishes for something, one imagines an end to be good. But this evaluative orientation proves to be a major limitation on the kinds of imaginings that could constitute wishing since what is good is determined in relation to animate things. One may wish the good for a friend (EN 1157b36), but one cannot wish the good of something without a psyche (τῶν ἀψύχων . . . οὐδὲ βούλησις ἐκείνῳ ἀγαθοῦ). For it "would be ridiculous perhaps to wish the good for wine, though one might, if at all, wish that it be preserved so that it is available to him" (EN 1155b28–31). For an inanimate thing such as wine to be good at all, it must be good for an animate being. The connection of imagination to wish is therefore conditioned by the fact that whenever one wishes, one imagines an end to be good for some living creature.

There is a social aspect to what one imagines to be good. People who live together tend to pursue similar activities, since they wish for and suppose that some activity is that in which living together consists. This is why there are communities dedicated to drinking, playing dice, and hunting, as well as a community of philosophers (EN 1172a3–8). The diversity of communities is explained by the diversity of ways in which people see the world, imagining different things as goals. Hence bad people suppose that they are pursuing noble aims, so that their "wishes are unclear," and those who are unjust go about life as if "what they wish to do is just" (EN 1178a30–31). Just like excellent agents who wish the good for themselves and their community suppose that they aim at the good, unjust people suppose their aims are good, even if they are aware that their suppositions are not endorsed by others.

Aristotle's view, as reconstructed here, has the peculiar consequence that one could not form a wish for universal suicide, the goal of which would be to eliminate all life in the universe.[10] To be a wish, such a desire would need to be rationalized as counting toward the good of someone. But perhaps this is not entirely surprising. To wish for an end, in Aristotle's view, is in part to take that end to be good for someone else. And, as argued previously, taking an end to be good for someone counts as both a reason for acting in a way that brings that end about and as a desire for the deed that brings it about to be done.

This view, on which to wish for something is to imagine it to be good for someone and to desire it, explains how one can fail to wish for the correct end. When one imagines falsely, a bad end appears to be good. So one wishes for it, supposes that it is desirable, and is motivated to act to achieve it. Consequently, "intellect is always correct but imagination is both correct and not correct. On account of this, the object of desire always moves, but this is the good or the apparent good" (DA 433a26–30). It is the coincidence of the actual good with the apparent good that is characteristic of an excellent agent. An excellent person "wishes and does what is both good and appears so" (EN 1166a15). This representational fidelity in the psyche is what makes one excellent. Only when appearance and reality coincide will one consistently form the right aims.

A good disposition is not an insensitivity to the passions nor to pleasure and pain but rather a propensity to feel these passions "at the right times, with reference to the right objects, toward the right people, with the right motive, and in the right way" (EN 1106b18–23, cf. 1105a1–9). Hence, the correctness of desire can be identified with loving the fine, a state that enables one to make good decisions and live well.[11] It is with respect to this state rather than with respect to passions that one is called good or bad (EN 1105b30–2). As Lännström puts it, "To be virtuous is to have a tendency to act, feel, desire, and choose correctly. In other words, to do so according to reason and for the sake of the fine. To have an individual virtue is to have a tendency to act, feel, desire, and choose correctly in a particular sphere."[12] To recognize what is fine as fine is antecedent to belief, since what appears fine in part informs the process by which one reaches beliefs. So one's conception of fine depends on how the world appears to one. Excellence is not a matter of having true beliefs but of getting appearances right.

According to this nondoxastic account of excellence, an excellent agent "wishes and does for himself what is both good and appears so" (καὶ βούλεται δὴ ἑαυτῷ τἀγαθὰ καὶ τὰ φαινόμενα καὶ πράττει) (EN 1166a13–18). Differences among agents' characters therefore correspond to what appears good to them. But an agent is the cause of what appears to himself because he is the cause of his own habits (EN 1114a31–b3). Failure to get this appearance right comes about when the imagination presents things inaccurately, paradigmatically when one is overcome by a passion: "If someone is corrupted because of pleasure or pain, the right principle (ἀρχή) cannot appear to him (οὐ φαίνεται), neither that he ought to do it, nor that he should choose and do everything for the sake of it" (EN 1140b17–19). Someone whose discernment is compromised by pleasure or pain, like one who is suffering from illness or sleepiness, will be apt to take what merely appears good to be actually good. Doing badly is typically a failure of imagination.[13]

This interpretation of Aristotle's psychology of desire suggests that the excellent individual's psyche will be characterized by an agreement of imagination and intellect. Indeed, Aristotle takes the agent whose imagination and intellect agree to be a model of how one ought to deliberate and act. "The excellent agent (ὁ σπουδαῖος) discerns things rightly, and in each case the truth appears to him (ἐν ἑκάστοις τἀληθὲς αὐτῷ φαίνεται). For to each particular disposition, there are properly fine and pleasant things, and perhaps an excellent agent is distinguished most by seeing the truth in each case, being like a standard or measure of others" (EN 1113a29–33). The tendency to have phantasmata in which what is really good appears as good is constitutive of what it is to be excellent. Indeed, Aristotle claims that even the phantasmata experienced by a good person during sleep are better than those experienced by ordinary people (EN 1102b10–11). The possibility of this agreement of imagination and intellect depends on the assumption that in wish, reason and desire are not opposed elements in a mechanism of choice but two distinct descriptions of an agent, which explain why he is motivated to pursue a certain aim. This account of moral success and failure is nondoxastic in the sense that bad wishes, rather than false beliefs, are taken to be what makes one act badly.

Aristotle accepts neither the realist nor the relativist horn of the dilemma concerning the apparent good. Instead, he dissolves the dilemma by means of the power of imagination while preserving the phenomena introduced in the puzzle. Like the realist, Aristotle holds that there is a proper object of wish, the good life. But Aristotle rejects the false premise of realism, that desire stems from one's beliefs. The imagination, and hence what appears to one, need not conform to one's judgments. Like the relativist, Aristotle holds that wish is for the apparent good. But he rejects the false premise of relativism, that there is no natural good for human beings. Since the imagination can agree with intellect, appearances can conform to what is naturally good. The invention of imagination thus supports a richer moral psychology than is countenanced by the realist or the relativist, undermining the very terms of the debate. Because it can produce and present appearances that are true or false, independently of the other powers of the psyche, the imagination provides the complexity necessary for explaining how both moral success and failure are possible. For Aristotle, one's goodness depends on the attunement of appearance and reality in one's psyche. But wishing does not make it so. Thus, practical reasoning, if it is to determine action, must not only render an end but also deal with the means necessary to realize that end.

CHAPTER 8

DELIBERATING

Deliberation (βούλευσις) is "not about ends but about what is for (πρός) the end" (EN 1112b11–12, cf. 1112b33–34). Choice (προαίρεσις) concerns what "has been deliberated upon" (EN 1113a4) and is thus also "about that which is for (πρός) an end" (EN 1111b27). A straightforward reading, in light of this shared "πρός," would take deliberation to be the process by which, starting with a set of possible means instrumental to achieving a goal one has in view, through the application of rules, one removes all but one choice. This remaining course of action, as what has been deliberated upon, would thus be chosen before the others.

Both the instrumental and deductive aspects of this interpretation have been disputed. Wiggins points out that "πρός" does not always refer to instrumental means but can indicate those conditions one takes to constitute the end one has in view.[1] Hence deliberation could not consist solely in examining alternative causal sequences but must involve consideration of constitutive means. Such noninstrumental reasoning is not reducible to calculation of the probability that some course of action will bring about what one wants. For it involves the consideration of whether various possible combinations of outcomes would count as achieving one's aims. Indeed, Wiggins hopes to dissociate "Aristotle's whole theory of deliberation from that pseudorationalistic irrationalism, insidiously propagated nowadays by technocratic persons, which holds that reason has nothing to do with the ends of human life, its only sphere being the efficient realization of specific goals in whose determination or modification argument plays no substantive part."[2] In practical, as opposed to technical, deliberation, one reasons about goals endogenous to one's conception of the good life rather than about exogenous outcomes.

Apart from this effort to dispel the instrumental interpretation of Aristotelian deliberation, it has also been noticed that practical reasoning has an ineliminable particularity, since a universal rule cannot be "stated to include all the circumstances of its application."[3] Indeed, McDowell has pointed out that the instrumental interpretation of deliberation falls into a regress of rules and applications.[4] If what one should do could be deduced from a universal rule, how could one know that this rule, rather than some other, is to be applied in a particular situation without appealing to another rule to determine its conditions of application? This regress of rules and applications is broken when one recognizes knowing what to do must consist in part in grasping a sensible particular.[5] Indeed, Aristotle insists that one can know a universal statement to be true but fail to act according to that principle, since "what is practicable are particulars" (EN 1147a3–4, cf. 1141b16–28). Practical wisdom itself is distinguished from science insofar as the former, but not the latter, concerns the particular (EN 1142a23–24)—that is, what is practicable (EN 1141b21). But while action always arises within a particular situation, deliberation is nevertheless a form of reasoning, which Aristotle characterizes as a kind of search.

Like geometrical analysis, deliberation works "back from the result aimed at."[6] In the geometrical case, what is aimed at is a completed geometrical figure, while what is sought is a progression of geometrical "actions" that would allow for the construction of that figure. In the practical case, what is sought is the natural, human good (EN 1141b8–9). As opposed to productive, technical endeavors such as carpentry, in which the product is something external to the activity that brings it about, in practical deliberation the "good action is itself the end" (EN 1140b7). Yet whether the product is a geometrical proof, a table, or a good action, the reasoning begins from a pre-given goal and seeks the means that are constitutive of that goal.

This characterization points to the hypothetical or provisional nature of deliberative reasoning. Just as one assumes in geometry that if the angles of a triangle are equal to two right angles, a desired conclusion follows, so also one hypothetically assumes an outcome to be achieved in productive sciences. For example, medicine assumes health as the desired outcome (EE 1227b25–32). A physician could heal or kill the patient. Yet in this example of technical deliberation, the physician does not need to argue that health is good but assumes it hypothetically. Since the deliberative part of the psyche contemplates the reason for an action as a cause, those who fail to lay down a fixed aim (σκοπός) do not even count as deliberators (EE 1226b30). Segvic notes that the hypothetical status of the ends or "principles" of action posited in deliberation are provisional but need not be tentative.[7] For example, one's wish might constitute a comprehensive conception of the good life. Yet, relative to deliberation, it would be provisionally given. For one might, after all, have deliberated about a different end. Aristotle therefore treats deliberation as taking something as provisionally given and, on that basis, seeking the constitutive or instrumental means required to bring about that aim.

Aristotle's comparison of deliberation to geometrical analysis reveals another similarity between them, which is sometimes lost in modern commentaries and translations. In contrast to a deductive proof, geometrical analysis is heuristic and ampliative rather than deductive. Working "backward" from what is set down to elements, analysis terminates in the presentation of a synthetic path that can only be achieved after the analysis is complete. Aristotle's most extensive description of deliberation draws on this nondeductive character of analysis, as if it were obvious to his auditors.

> We deliberate not about ends but about what is for (πρός) ends. For neither does a physician deliberate about *if* he should heal, nor an orator about *if* he should persuade, nor a statesman about *if* he should make good order, nor any of the others about their ends. Rather, positing (θέμενοι) the end, they consider (σκοποῦσι) *how* or *due to what* it might be. And if it appears that it might come about in many ways (καὶ διὰ πλειόνων μὲν φαινομένου γίνεσθαι), they consider which is easiest and best. But if it can be accomplished in just one way, then about how it can be done by this, and what else it requires, until he arrives at the first cause, which is the last in the order of discovery (ὁ ἐν τῇ εὑρέσει ἔσχαστόν). The deliberator is like one who seeks (ζητεῖν) and analyzes (ἀναλύειν) in a manner as if by a diagram (τρόπον ὥσπερ διάγραμμα). But it appears that not all seeking is deliberation, such as mathematical inquiries, but all deliberation is inquiry; and the last in the analysis is the first in the coming to be (ἐν τῇ γενέσει). (EN 1112b11–24)

Deliberation and geometrical analysis are species of seeking. Both begin by positing something hypothetically, a completed end. Then they work "backward" from this posited result, asking what means might bring it about. They are finished when they hit on the first step by which the genesis, of the figure or of the practical end, might come about.[8] Both geometrical analysis and deliberation are instances not of deduction but of ampliative reasoning. And both are contingent. One might hit on a satisfactory way of bringing about the end without exhausting every option, or one's search might fail.

Aristotle's analogy between geometrical analysis and deliberation highlights their shared structure. Analysis is apt to increase geometrical knowledge by providing the geometer with an arsenal of constructive capabilities that may be called on in formulating deductive proofs. Deliberation is, similarly, apt to increase practical knowledge, since after one has deliberated, one knows what to do in a given situation. When there is just one way of producing the end, the deliberator is in a position analogous to the geometer who is analyzing what Amphinomus called an "ordered" problem. Asking how or due to what the posited end might come results in a unique solution. But if there are finite but multiple ways to solve it, it corresponds to an "intermediate" problem, and one must decide which solution is easiest or best.

The shared, ampliative aspect of these forms of reasoning suggests that they also share a cognitive basis. In the *Ethics*, Aristotle avoids a direct discussion of this cognition. Nevertheless, he makes it clear at least that to discover the best or easiest solution, one must consider what appears to bring about the end. This suggests that the imagination's capacity to produce and present a variety of different, even mutually exclusive options to be reasoned about is required for deliberation. It is therefore rather unsurprising that, in his psychology, Aristotle describes a kind of cognitive capacity or activity that he gives the peculiar name of "deliberative imagination."

"All imagination," says Aristotle in one of the last chapters of *De Anima*, "is either calculative (λογιστική) or perceptual (αἰσθητική)" (DA 433b28–29). The designation of this power of the psyche as calculative suggests that it participates in reason. And in the *Ethics*, Aristotle characterizes the part of the psyche that calculates as being generically the same as that which deliberates (EN 1139a4–15). Yet Aristotle was able to individuate the imagination as a distinctive part of the psyche only by establishing a boundary between it and judgment. How can imagination be involved in calculation, a paradigmatically reasoning activity?

Aristotle describes deliberative imagination by contrasting it with the perceptual imagination possessed by beasts incapable of deliberation. Can "indwelling imagination" be attributed even to "incomplete" animals, that only possess the sense of touch? Yes, since appetite is attributed to such animals, imagination must be as well.

> For pleasure and pain appear (φαίνεται) in them, and if there is this, there is also necessarily appetite. But how could there be imagination for them? Or is it that, just as they are moved indeterminately (ἀορίστως), so also these things are in them but in them indeterminately? So perceptual imagination (ἡ . . . αἰσθητικὴ φαντασία), as it was said, is that which is attributed to other animals, but the deliberative (βουλευτικὴ) is in the calculative animals (ἐν τοῖς λογιστθκοῖς) (DA 434a2–7). For whether to do this or that is already the function of calculation. And it is necessary to measure by one thing. For it pursues the better. Hence it can do just one deed from many phantasmata (ἓν ἐκ πλειόνων φαντασμάτων ποιεῖν). (DA 434a5–10)

An animal's self-motion is linked to its participation in the imagination. Incomplete animals are "without ends" and move indeterminately in virtue of their possessing imagination indeterminately.[9] Appetite depends on the appearance of something as pleasurable or painful and hence requires the activity of imagination. But an incomplete animal, insofar as it perceives the object of desire through direct contact, is in some sense deficient in ends.

Could such an animal be aware of an object of desire but be unaware *that* it desires that object? If so, an absent object of desire would be presented to the animal possessing only perceptual imagination. But its presence to the animal's psyche would not give rise to a reflective attitude or self-aware, conscious state. This is a way of understanding how imagination might be in an animal only "incompletely." An objection to this is that if Aristotle holds that everything done by desire is voluntary, then the perceptual imagination is sufficient to determine voluntary actions.[10] Thus, even an incomplete animal would be aware of its desire for an object. In that case, what distinguishes perceptual imagination might be that it offers an insufficient basis for weighing alternative courses of action against one another. An animal might be aware that it desires an object and thus acts voluntarily in pursuing it but would be unable to envision alternative ways of satisfying that desire.

On either reading, the imagination plays a definite role in desiring. Imagination is a power in virtue of which an animal is "activated toward something" and moves, and can give its movements a final cause (MA 701a5–30). All sublunary animals participate in the nutritive psyche and are directed toward their own survival (DA 413a23–28), and all animals in addition can pursue goals in virtue of their desires (DA 432b21–27). The indeterminacy of animal motion, then, might be analogous to the incompleteness of plants' participation in touch. Though plants lack the perceptual capacity to receive the form of touch without the matter (DA 424a32–b3), they are affected by compounds of form and matter impressed upon them. In Jonas's terms, plants and animals have a different "mode of exteriority" than plants, illustrated with the concept "indeterminate affections."[11] For example, when air is filled with an odor, it is indeterminately affected, since the perceptible form does not remain when the matter is removed (DA 424b15). Analogously, indeterminate animals have a mode of interiority in which their actions are dependent on only partial appearances of objects of desire and thus on only partially formed aims. Such animals lack a single experience.

Brutes cannot be unruled, since they have no universal beliefs to be resisted by appetite and only the "appearance and memory of particulars" (EN 1147b4). But in human beings, many memories of the same thing "bring to completion" the capacity for a "single experience" (*Met.* 980b21). Human actions are thus guided by an awareness of one's own aims and are calculative, participating fully in imagination.[12] Deliberative imagination is distinguished from perceptual imagination in terms of calculation, which depends on an agent's having a single measure from which a calculation can be made. Yet Aristotle does not say that the act of measurement, which is doxastic and discriminative, is itself the work of the imagination. Thus, while commentators sometimes maintain that the imagination is responsible for synthesizing phantasmata into a single standard, that cannot be what Aristotle means here.[13] Rather, Aristotle's point seems to be that the imagination is responsible for producing and presenting one with a variety of courses of action that potentially satisfy one's aim. Each course of action presented in a phantasma is then subjected to rule, which is given by the power of reason or calculation. This results in a process of elimination, by which each phantasma is weighed and eliminated if it fails to satisfy the standard. What is produced out of many phantasmata is not the rule but a single course of action. This is the choice of a deed.

According to this interpretation, Aristotle's conception of deliberative imagination can support both sorts of deliberation outlined in his *Ethics*. If there is only one way to satisfy one's end, as corresponds to the "ordered" construction problem, the deliberator must reason as follows: If my end is Z, and a course of action Y is presented in this phantasma, does Y lead to Z? If not, eliminate Y. If so, what actions might lead to Y? This process is then iterated until the deliberator finds a route to Z through a series of mediating actions. The deliberation is done when one finds an immediately practicable action A, from which, by performing the actions in reverse order, Z eventuates. Here the stage corresponding to the geometrical synthesis would be the undertaking of the whole series of actions, from A to Z. When there are multiple but finite courses of action, so that the search corresponds to the analysis of an "intermediate" problem, the deliberation is similar. But in this case, the deliberator will not just need to ask whether Y could bring about Z but also whether an alternative course of action X would be better or easier than Y. Thus, the "measure" or "standard" employed in each case is different. In the first kind of deliberation, it is solely a satisfaction criterion. In the second case, the standard must also determine which of two or more courses of action that would satisfy Z is preferable.

This interpretation of deliberative imagination as the cognitive function underlying the production and presentation of courses of action as satisfying or not satisfying is a rule corroborated by some of Aristotle's other stated views. In the *Ethics* account, which does not make direct reference to all the powers described in his psychology, Aristotle holds that the calculative part of the psyche has reason (EN 1139a14–15) and characterizes deliberation as being excellent just when one has a correct reason (EN 1139a24–5). Thus, while deliberation involves actively envisaging which resources are at one's disposal, choice comes about when one determines, on the basis of one's suppositions, which course of action is most likely to satisfy one's wish. This activity is calculative and explicitly doxastic, reducing the multitude of potential combinations of ends and means to a single course of action that satisfies one's end.

This cognitive side of this process is not, of course, described in any detail in Aristotle's ethical works. But there is a passage in *De Anima* where he refers in some detail, albeit obliquely, to this function of the imagination.

> The intellectual part (τὸ νοητικὸν) then thinks (νοεῖ) the forms (εἴδη) in phantasmata; and as what it should pursue or avoid is defined (ὥρισται) in these, it is moved even in the absence of perception, when phantasmata are before it. For example, perceiving a beacon that is aflame, from the sight of it moving by the common sense, one comes to know that it is hostile. But it is by the phantasmata and thoughts in the psyche that, as though by sight, one calculates (λογίζεται) and deliberates (βουλεύεται) about what to do with respect to what is present (τὰ παρόντα). And when it pronounces that something holds what is pleasant or painful, then it pursues or avoids it—and in general these will be one action (DA 431b2–10).

This passage illustrates the work of deliberative imagination. To decide what to do in the situation under consideration, the agent must calculate and plan, an activity that occurs in phantasmata. But this process is also intellectual. As the imagination runs through various options, the intellectual part reduces them to a single choice.

Aristotle describes a wartime scenario in which one is trying to plan whether to pursue or avoid a beacon in the distance. Perception of the beacon could not alone determine the correct course of action, since the deliberating agent does not know whether it belongs to an enemy or to an ally. One must imagine a number of different scenarios from the way the beacon moves and from other aspects of its appearance. Is there reason to think, from the way it moves, that it belongs to the enemy? If so, what is their approximate number? Is the beacon a signal of distress? If so, is there reason to believe that there are reinforcements in the area? Situation-specific questions balloon at this point. Being able to answer these questions depends on one's capacity to imagine what is absent to perception as well as to represent what one knows about one's own resources. How many soldiers and horses does one have at one's disposal? Making a choice depends on considering these various appearances and then calculating, on the basis of one's rational standard, whether it is prudent to pursue or avoid the beacon.

Choice (προαίρεσις) is a compound of προ- and αἵρεσις, so that, Aristotle comments, "the name itself seems to suggest that it is something taken before other things" (EN 1112a15). This is sometimes taken to mean Aristotle considers that one course of action is chosen "before" others in a preferential sense, so that one prefers that action to others. Yet, Gisela Striker informs me that "before" more typically has a temporal sense, so choice indicates a decision to commit to undertake a course of action before the action itself occurs.

Aristotle holds that practical reasoning and theoretical reasoning are distinct (DA 432b27–29). But this division does not map smoothly onto that between particulars and universals. For example, the young can become good mathematicians but not physicists or statesmen, since mathematical things are known "on account of abstraction" while the principles of natural science and practical philosophy are known from experience. Thus, the young "do not become convinced" about empirical or practical things but only speak of them. In distinction, which judgment to make about particulars "is not unclear for those with experience." The young cannot reliably reason about these things because they have not encountered a sufficient number of comparable, particular situations. Similarly, an agent may go wrong in practical reasoning concerning either "that all heavy water is bad or that this particular water is heavy" (EN 1142a17–23). Aristotle here groups natural science with practical wisdom, since both require knowledge of particulars gained through experience, and consequently require "a lot of time" (EN 1142a14–16). While a train of geometrical reasoning goes wrong only when one misapplies universal concepts, practical reasoning has an additional source of error in the misidentification of particulars for which experience is a precondition (AnPr. 46a17–27, AnPo. 100a3–8).

These comments, and the discussion of practical syllogisms that stems from them, have led readers to worry that in the domain of practical reasoning, Aristotle conflates the scientific knowledge of universals with particular knowledge gained by experience. A related concern is that the practical syllogism seems to describe the kind of rational, practical reasoning of deliberating agents but also to characterize the purposive activities of nonrational animals. Anscombe gets around these problems by arguing that one need not be aware of the syllogism one performs.[14] But Striker has provided an even more attractive solution, with the argument that the syllogism simply does not describe deliberation. Deliberation is, after all, a search rather than a deduction. Thus, the practical syllogism only describes what happens after the deliberation has been completed. In this view, choice, the product of deliberation, provides the first premise in the practical syllogism. But the second premise must come from perception or experience. The practical syllogism is a description of what happens after a policy has been chosen, not of the deliberative activity itself.[15]

Choice, as the product of deliberation, is the terminus of a process in which one considers a number of prospective situations. As in geometry, in deliberation one must think hypothetically, producing and presenting oneself with absent objects. But although they spring from the same cognitive soil of the imagination working in the service of reason, deliberation and analysis have fundamentally different consequences. Successful analysis produces a number of steps from elements that, when synthesized, generate a geometrical construction. Generative language is necessary to describe the geometer's procedure, and the synthesis of the figure may be imagined as unfolding temporally. But insofar as geometry is a purely theoretical endeavor, the steps in this generation are not really separated in time. But whereas geometrical seeking ends with a proof, the last step in the deliberator's analysis "is the first in the genesis" of an action (EN 1112b20–24). When one deliberates, this "first step" can be understood as a choice, the major premise of a practical syllogism. But since the world in which one acts is constantly changing, seeking the best way to reach one's goal could never have the kind of closed, deductive structure afforded by mathematics. Instead, the best way of achieving one's end—the choice—might itself need to be revised, depending on the circumstances.

The relation between deliberation and action can be made more precise by considering the peculiar modal, temporal, and self-referential structure Aristotle attributes to practical reasoning. Theoretical reasoning concerns unchanging things, such as the incommensurability of the diagonal with the sides of a square, or changes that always happen in the same way, such as sunrises and solstices. But the practicable good must admit to being other than it is (DA 433a29–30).[16]

In addition to this modal constraint on deliberation, it has a particular temporal structure of being prospective. One cannot deliberate about the past, since "the chosen is not what has become, so no one chooses that Ilium be destroyed. Nor would one deliberate about what has come to be; but rather about that which is capable of coming to be and also of not coming to be" (EN 1139b5–9). Past events are not real objects of choice. However, deliberation also concerns the present since "one calculates and deliberates about what one is about to do from what is present" (DA 431b7–8). What is "present" has both a spatial and temporal significance. But "what one is about to do" could only indicate a possible future event. Thus, deliberative activity must conform both to modal and temporal criteria, insofar as it concerns only possible future states of affairs but connects those possibilities to the concrete situation presented in perception.

Aside from these modal and temporal constraints, in deliberation one must believe that one is oneself able to bring about the future state in question. Only a madman or fool would deliberate about Hades or whether the diagonal is commensurable with the side of the square. Thus, one who has one's intellect will only deliberate about those things one can do (EN 1112a18–31). The capacity to distinguish what is up to us from what is not up to us is intellectual. Only one who had lost hold of intellect would deliberate about something that could not be achieved. This point is made even more forcefully in Aristotle's *Rhetoric* (I.4), where he maintains that one must take the action about which one deliberates to be "up to us" and to believe oneself to be the "first principle" of its genesis (ἡ ἀρχὴ τῆς γενέσεως). To deliberate rightly, one must have not only a belief about the contingency of the situation one plans to change but also a de se belief concerning one's own capacity for action that will effect that change.

This de se belief must consist in part in a recognition that one is able to choose between contraries. Rational powers, as opposed to the capacities of nonrational animals, "are productive of contrary effects" (*Met.* 1048a2–5).[17] Thus, there must be something governing the rational powers, driving them to produce one or another of the contraries. What governs this power in the case of action is desire or choice (*Met.* 1048a10–

11). Such thinking activity is "complete" and comes to be spontaneously when the conditions that could bring it about are satisfied—that is, when it is desired when nothing external prevents it (*Met.* 1049a5–9). Genesis, or coming to be, is not a gradual movement toward an object but a complete and instantaneous change (GC 317a18–27). Thus, while discursive thought as such "moves nothing," it nevertheless moves something "as that which is for the sake of something and is practical" (EN 1139a35–36). Choice is therefore a form of thinking and a motive force, in Aristotle's locution, "desiring intellect" or a "reasoning desire" (EN 1139b4–5). Consequently, no step is necessary to mediate between the end of sequence of practical thinking and the beginning of the action.

There may, of course, be an intermediate step, even a long lag, between deliberation and acting on one's choice.[18] A choice may be made, a plan laid, a policy selected, without its ever being enacted. But Aristotle need not posit any intermediate step between calculating and moving. Deliberation is constrained to contingent, future states that one correctly believes are in one's power to bring about or prevent, and practical reasoning itself can give rise to action. This metaphysical description of the relation between thinking about an end and bringing it about is underwritten by the future-perfect, contingent form of the phantasmata from which the deliberator searches for present states of affairs that might make them true. Deliberation is thus akin to geometrical analysis but distinct from such theoretical searching in that it is apt to bring about not only imagined but real change.

CHAPTER 9

SHAME *and the* OTHER

Aristotle's psychological doctrine of imagination, though motivated by theoretical problems concerning geometry, allows for a critical reevaluation of prior conceptions of cognition, motivation, and practical reasoning. In each of these domains, Aristotle uses the idea of a psychological power to produce and envision absent, hypothetical, or counterfactual states of affairs to offer novel interpretations and explanations of quotidian cognitive activities. Insofar as it is necessary for thinking, the imagination must be invoked in any complete description of the rational activities of wishing and practical reasoning. What is more surprising is that Aristotle uses the imagination not only to explain such rational activities but also in his account of the passions (πάθη).[1] In this chapter, I examine how this strategy allows Aristotle to connect the social, psychological, and bodily aspects of one particular passion: shame.

Shame is important for my purposes not only because his analysis of this emotion appeals to the imagination but also because shame is integral to Aristotle's conception of how agents form wishes and pursue plans. This book has argued that for Aristotle the object of every wish appears to be good for an irrational animal, a member of one's community, or oneself. But an agent need not have a comprehensive theory of what it is to thrive as an irrational animal, community, or human being to have some end in view. All that is required is a conception of the end of the creature for whom the agent wishes good, a conception that forms the content of the supposition on which the agent acts. This conception must be evaluative, but it could be so in a "thin" sense, such as in the evaluation that "it is good to hunt" or "it is good to drink." Such a conception of how to live, though insufficient to constitute a "thick" philosophical account of the good life, is sufficient to support a wish, and hence to motivate rational deliberation about how to bring about the desired state of affairs specified in that wish. Differences in wishes depend fundamentally on different ways in which agents see the world.

How one sees the world is constituted by one's upbringing and habituation to within a specific society. Aristotle thus maintains that the legislator must understand enough psychology to influence citizens to wish for and pursue fine ends while avoiding bad ones. Consequently, the legislator must distinguish between voluntary and involuntary actions, since he is concerned with giving out honors and punishments (EN 1109a30–35, 1102a23–25). Actions or passions are said to be excellent only insofar as they are voluntary, since the involuntary are done by force or through ignorance (EN 1109b35–1110a1).[2] Effective legislation therefore depends on an ethical understanding of fine and just things, which are the subject of political inquiry (EN 1094b10–11). Yet unlike the philosopher who inquires for the sake of knowledge, the legislator's goals are practical and concrete. Since every art is concerned with bringing something else into existence (EN 1140a10–14) and the goal of the lawgiver is to make citizens excellent (EN 1180b23–25), the good legislator uses political science to establish good laws. In this way, the legislator is analogous to an artist, who uses scientific knowledge to produce paintings (EN 1181a23–b2).

This distinction between theoretical science and the art of legislation helps explain Aristotle's view that politics is concerned with voluntary action. Agents are blamed when they avoid pain or do shameful things, since we anticipate these outcomes when we make a choice (EN 1110a31–b1). While children and animals act voluntarily (EN 1111a24–26), they escape blame because they do not deliberate about alternative outcomes and pursue definite policies, and thus cannot be said to act "by choice" (EN 1111b8–10). Nevertheless, it would be "ridiculous" to claim that adults motivated by irrational desires act involuntarily (EN 1111a27–31), since agents able to plan their actions are to blame whenever the end achieved is shameful (EN 1111a28). Blameworthiness is not determined by an analysis of action's etiology in a certain form of desire but rather by whether the agent could have foreseen whether the effect of her action would be shameful.

Shame is essential to Aristotle's conception of blameworthiness. Shame "concerns voluntary actions, but a gentleman (ὁ ἐπιεικὴς) would never voluntarily do something base" (EN1128b28–29). The term "gentleman," which traditionally connotes a person of the upper class, in a more general sense indicates one who is capable or reasonable. Reasonableness, in this context, is a kind of thoughtfulness that prevents one from doing something base. To avoid such base actions, one must be able to think prospectively about what actions would be considered shameful by others. Thus, "shame would be *ex hypothesi* good, for if one were to act basely, then there would be shame" (EN1128b29–30). When an agent reflectively anticipates the likely effects of his action, shame acts as a "hypothetical good" in that it gives a criterion for choice. A sense of shame may be insufficient to decide the best course of action in a given situation, but it eliminates base actions from consideration. Hence Aristotle distinguishes two kinds of courage—one based on shame and the other on fear. The better kind comes to be "by shame (δι' αἰδῶ) and the desire for the fine, for this is honorable, and by fleeing reproach, which is shameful" (EN1116a28–29). The avoidance of shame, while not equated with virtue, is the negative activity corresponding to the love of the fine. Whether one's action is motivated by irrational or rational desire, one's expectation about whether the outcome of that action will be shameful is a criterion of blameworthiness. But if one's sense of shame acts to determine the norms governing actions, what is shame, and what gives it its particularly social and political functions?

The sociality of shame was recognized long before Aristotle, and shame appears in an explicitly public context in Homer's *Iliad.* When Thersites speaks against Agamemnon, criticizing the king for keeping the Achaeans in an endless war and for the abduction of the daughter of Apollo's priest, Odysseus responds that he has no right to speak against kings and, before beating him, promises to "uncover his shame."[3] From the description of the action that follows, "shame" (αἰδῶ) here probably means "nakedness" or "genitalia"—that is, something bodily that is uncovered and revealed publicly in the assembly of the Achaeans.

The political implication of the Homeric passage is that it is shameful for the rabble to speak against kings, but the political function of shame gets expanded in the subsequent literary tradition. Plato's *Protagoras* tells of a time at which humans already had language, religion, and the rudiments of cultivation but engaged in endless armed conflicts. To save humanity from destruction, the king of the gods sends Hermes to distribute the art of shame and justice among men. This art should not, Zeus says, be distributed only to a few specialist practitioners as the other arts were but to everyone: "For cities would never come to be if only a few possessed these, as is the case with the other arts. And establish this law as coming from me: Death to him who cannot partake of shame and justice, for he is a pestilence to the city" (*Protag.* 322a–d). In the Platonic myth, shame is a principle of justice and a necessary condition for political organization. Shameless people must be politically excluded because to act shamelessly is to do damage to the institution of the city as such. It is insofar as one sees oneself as seen, and hence as subject to the judgment of others, that one is constrained by the norms of the community. To be subject to the norms and laws of the city is to participate in shame and justice.

Following Plato, Aristotle holds having a sense of shame as necessary for political participation and recommends the banishment of citizens who are unable to be reformed through rewards and punishments (EN 1180a10). But if Plato had already interpreted the sense of shame as constitutive of justice in the political community, Aristotle's innovation is to analyze shame in terms of his theory of imagination. In applying this theory, Aristotle retains the political aspects of shame but reinterprets it as a complex bodily and psychological phenomenon dependent upon an internal appearance. According to Aristotle, everything that comes to be in the psyche must be either a passion, a power, or a state (EN 1105b20). Passions are "accompanied by pleasure or pain" (EN 1105b23), and shame is explicitly listed among them in the *Eudemian Ethics* (1220b11–14). However, shame is left out of the enumeration of passions in the *Nicomachean Ethics*. Moreover, in the *Eudemian Ethics*, Aristotle calls his tripartite distinction a "division," while the *Nicomachean Ethics* merely lays down the distinction, saying that the parts of the psyche "come to be three." Thus, the *Eudemian* language implies a real division, while the *Nicomachean* suggests there is a merely nominal distinction among these parts of the psyche. In both works, passions are taken to be hedonic, but in the *Eudemian Ethics* alone are the pleasure and pains involved in having an emotion called "perceptual" and "intrinsic." In other words, the *Eudemian* account stresses the embodied character of the passions.

While Aristotle's treatment of the shame in his ethical works do not match in all details, each of them is consistent with his most comprehensive account of shame, given in the *Rhetoric*. There Aristotle defines shame (αἰσχύνη) as some pain (λύπη) and disturbance (ταραχὴ) about the appearances of evils in virtue of which one bears disrepute (τὰ εἰς ἀδοξίαν φαινόμενα φέρειν τῶν κακῶν), whether present, already having become, or expected (*Rhet.* 1383b12–15). Following Grimaldi, I take the appearance to be of evils.[4] But it is not clear whether the apparent evils are the causes of *adoxia*, for example, bad actions on the part of the shamed party, or the consequences of *adoxia*, for instance, not having the protection of one's neighbors.

This tension can be resolved, I think, by noting that a feeling of shame is relatively independent of the time at which one acts. One can feel shame contemporaneously with a shameful act but also after one has done it or before one has acted. Disrepute or *adoxia*, conceived as bad reputation, stems from the opinions of others rather than from one's own judgment. Regardless of one's own assessment of the action, if what one wished to accomplish were believed to be dishonorable by relevant parties, one would feel the pain of shame. Konstan notes that shame "is not conceived as a response to repute or disgrace (ἀδοξία) as such, but rather to those ills that lead to such a state."[5] In other words, the feeling of shame arises from considering the evils with which one would be identified from the perspective of another person. The evil about which one feels shame thus might be the action itself or the consequences of the action in virtue of which one is held in disrepute.

Throughout his discussion of shame in the *Rhetoric*, Aristotle assumes that the passion is concerned with the good and bad and thus can motivate activities of pursuit and avoidance. Aristotle usually reserves such motivational language for desire but here extends the capacity to motivate to passions of the psyche that are not *stricto sensu* desires.[6] This motivational use of shame is underwritten by the notion that to be ashamed at something is to see it as bad, and thus to evaluate it. But what precisely does shame evaluate? In the *Eudemian Ethics*, Aristotle treats shame as a middle state between not heeding the opinions of others and paying them too much attention, concluding that one who has this sense will only consider the opinion of "those who appear reasonable" (EE 1233b27–29). In other words, shame involves an evaluation not just of one's own actions but of the character of others who could have an opinion about those actions. In this last respect, shame depends on seeing others as having undertaken doxastic commitments. Though one might feel shame without forming one's own doxastic commitment about the action in question, one must be able to predict or "see" how someone else would form a doxastic commitment when presented with one's completed act.

I have been arguing that having the capacity to experience shame depends on being able—albeit incompletely and often inaccurately—to inhabit another person's perspective, to envision what the other person would be likely to believe given her doxastic commitments. It is necessary, in order to make this kind of evaluation, for both the agent and his imagined observers to have the capacity to grasp reason. Hence one cannot experience shame in front of animals and children, agents without political reason (*Rhet.* 1384b24). Aristotle stresses this rational and political aspect of shame in the *Eudemian Ethics* as well, arguing that while courage can be motivated by experience and knowledge, inexperience and ignorance, intoxication, or irrational passions such as temper or love, only a sense of shame is said to motivate political courage (EE 1229a12–25). As opposed to the other species of courage, political courage is due to the convention of a political community (EE 1229a29–30). And to be political is to participate in a kind of reason or form of speech that is concerned with the affairs of a city.[7]

Given that the capacity to grasp reason and to attribute reason to others is constitutive of shame, it is unsurprising that scholarship on Aristotelian passions has examined their cognitive aspect. Nussbaum holds that emotions for Aristotle are necessarily about something, and that even irrational appetites are directed at an apparent good.[8] This entails that the passions are at least in principle corrigible, insofar as they are susceptible to rational evaluation. Sherman agrees, describing Aristotle's passions as modes of attention, and claims in addition that they are fundamentally intentional states in virtue of which one tracks morally relevant information and communicates it to others.[9] On Sherman's account passions are defined by their cognitive content, and it is in virtue of this content that one evaluates what is important. Indeed both the evaluative aspect of passion and its responsiveness to reason can be discerned in Aristotle's numerous descriptions of shame.

In the spirit of Nussbaum and Sherman, I suggest that feeling Aristotelian shame requires the employment of those rational cognitive capacities in virtue of which one attributes doxastic commitments to others. However, merely attributing opinions to others is necessary but insufficient for holding that those opinions are important. In order for the disapproval of a peer to affect one's shame, one must accord respect to that peer. When an agent does accord such respect, the peer's opinion can give pleasure or pain, a feature that distinguishes emotional experiences from purely cognitive judgments concerning others' beliefs. But how could the agent who feels shame become aware of this alien doxastic state and "see" his disrepute in the eyes of others?

Aristotle holds that shame is a *phantasia about adoxia* for its own sake and not for its results (*Rhet.* 1384a23–24). Konstan translates phantasia as "perception" or "impression,"[10] while Grimaldi calls it a "vivid impression."[11] However, I think that phantasia must, at least in this definition of shame, denote not a perceptual state but the power in virtue of which the affection of shame comes about. This experience arises from seeing one's own aims as bad from the perspective of another person, a "seeing" that is clearly not perceptual in nature. Aristotle thus interprets the proverb that "shame is in the eyes" not literally to say that shame is perceptually given but rather to mean that one is most ashamed of what is "apparent" (*Rhet.* 1384a34–b2).

The doctrine that shame arises not from just any bad opinion others might have of one but from opinions one respects explains why the phantasia invoked in Aristotle's definition of shame cannot be perceptual appearance. One feels the pain of shame only from imagining one's *adoxia* in the eyes of those by which one wishes to be admired (*Rhet.* 1384a27–31). Consequently, even if that party were present, the appearance that gives rise to shame would be in part constituted by one's own previous commitments, expectations, and hopes concerning the opinions of others.[12] Likewise, one is not ashamed by every opinion but only by the opinions of those one respects (*Rhet.* 1384a25–26). One would not, for example, feel shame for violating the principles of a group of religious fanatics unless one were also a member of that group. Far from describing a vivid or faint perceptual impression, "phantasia" in shame indicates the imaginative presentation of one's disgrace in the eyes of a specific person or group that one has previously judged to be worthy of consideration.

This portrayal of shame as emerging from the act of imagining one's disrepute among those one respects gives shame a definite normative function. Just as the object of wish is an apparent good, the object of shame is an apparent evil. Consequently, one's sense of shame is regulative of the kinds of wishes one forms and acts upon. The avoidance of shame will tend to drive an agent to form goals that accord with those of his peers. In degenerate political communities, where good actions are generally held in disrepute, this will have the effect of reinforcing detrimental norms. But in thriving political communities, the cultivation of a sense of shame among citizens will induce them to uphold the laws, and thus to cultivate the virtues. Hence although Aristotle holds that one is not said to be good or bad with respect to passions but rather with respect to virtues and vices (EN 1105b30–32), it is clear that passions play a regulative role in determining the actions one undertakes. The feeling of shame is not intrinsically good or bad for Aristotle, but insofar as it tends to contribute to a shared sense of what is right, a sense of shame binds an agent's goals and plans to those of his community. Cognitively, this affective binding depends on the power of imagination to put together disparate and absent objects—one's imagined action and others' imagined responses to it—in a single presentation.

Leighton has argued that Aristotelian emotions may be aimless, so that one could feel shame about nothing in particular.[13] But if being ashamed depends on imagining one's status in the eyes of others, the object of shame must be a disgraceful action. In exercising the power of imagination, one produces a phantasm of that disgraceful act, toward which one's shame is directed. It is on the basis of this intentionality that Aristotle distinguishes between appropriate and inappropriate emotional responses. Moderation in shame is the tendency not to feel ashamed about everything or about nothing at all, but only about those things that really would meet with disapproval from the persons one holds in high regard (EN 1108a31–36). To have such a sense of shame is not itself a virtue, but it is praiseworthy insofar as it amounts to a propensity to regulate one's wishes and actions according to the expectations of the most admirable people one knows.

I have been arguing for two related theses. First, the capacity to present something absent to oneself by imagining it causes the feeling of shame. Understanding Aristotelian shame thus depends on appreciating how the ashamed person's imagining of the opinions of others is explanatory of the pain he feels. Second, this capacity to imagine the opinions of others is praiseworthy when it tracks, in a way that is neither overly sensitive nor insensitive, the opinions of the people by whom one wishes to be admired. When it does so, a sense of shame gives criteria either for revising one's wishes, when satisfying the end one wishes for would lead to a shameful outcome, or for revising one's plans, when some step of the plan would be seen as shameful. While there is nothing virtuous about the experience of shame as such, one's sense of shame can therefore be regulative of virtuous action.

Given that shame is a passion that depends on having an internal appearance of others passing judgment on one's character or action, it might seem that Aristotle conceives passions in a Cartesian sense, as properties of the mind rather than of the body.[14] However, Aristotle's account of passion is non-Cartesian in its insistence that a passion consists in or must be accompanied by pleasures or pains. This is evident, for example, in Aristotle's description of shame in the *Rhetoric* as a "pain" or "disturbance."

Leighton has argued that although all the emotions in the *Rhetoric* are described as being accompanied by pleasure and pain, this is not necessary for the treatment of passions in *De Anima*.[15] Fortenbaugh takes note of Aristotle's comment that the passions are "for the most part" accompanied by "sensory pleasure and pain" (*EE* 1220b12–14) to offer an alternative account. Fortenbaugh writes that Aristotle adds the qualification that such pleasure and pain is "sensory" because Aristotle "is referring to pleasant and painful sensations caused by bodily changes."[16] Fortenbaugh's interpretation is that pleasure and pain properly accompany most emotional episodes, though they might in specific instances not be felt, and hence a passion could be experienced without one noticing the corresponding bodily affection. Nevertheless, Fortenbaugh remains committed to the thesis that thought is the efficient cause of passion.[17] I think Fortenbaugh's interpretation is persuasive, with one caveat. In my view, the body is the efficient cause of the pleasure or pain experienced during an episode of shame.

According to Aristotle, when considering any passion in an animal, two different but correct descriptions are possible: one that makes reference only to the psyche and one that makes reference to the body alone. "The natural philosopher and the dialectician would give a different definition of each of the passions—for instance, the question "What is anger?" For the dialectician will say that it is a desire to pay back with pain or something like it, while the natural philosopher will say that it is a boiling of blood and hot stuff around the heart. And of these one will be giving the matter, the other only the form and account" (DA 403a30–b3). This does not answer the question of whether passions are psychological events caused by the body, bodily events caused by the psyche, or compounds of the two. Rather it establishes that a passion can be given two different, but seemingly equally legitimate, descriptions.[18] The dialectician describes the psychological processes by which the physical affection comes about, while the physiologist describes the physical product. But despite this de dicto separability, whenever one suffers a passion, "the body suffers (πάσχει) something" (DA 403a18–19). Passions are ex hypothesi common to the soul and body (DA 403a10–12), and are de dicto but not de re separable from the affections the body undergoes when one feels them.[19]

Aristotle's descriptions of the passions frustrate the Cartesian impulse to locate them on one side or the other of a soul-body division. His guiding observation is that "it seems that all the passions of the psyche are with body" (DA 403a16–17). Passions thus belong to and are properties of the psyche. Yet insofar as they are "not without" body, they are constituted by or emergent from a bodily state. Employing hylomorphic language, Aristotle considers passions as analogous to the form of straightness, which never appears without some body in which it inheres—that is, without something's being straight. Applied to the case of shame, Aristotle's suggestion is that although one might define shame without making reference to the body (as in the *Rhetoric*), in every actual episode of shame, the psychological and bodily activity of suffering shame will coincide.

The comparison of the imprint of psychological shame upon the body and the imprint of the form of straightness on a piece of metal is, however, imperfect. A straight piece of metal just is the form of straightness imprinted in metal. Every straight thing participates in this form, and nothing that lacks this form could be straight. However, as noticed by ancient readers, it seems that the case of the passions is more complex, since any bodily state associated with a passion could arise from other causes. Commenting on Aristotle's characterization of shame, for instance, Porphyry argues that the paradigmatic bodily state of shame—the blush—could be caused by other bodily dispositions, without the subject experiencing the corresponding passion. "Just as turning red from shame or turning pale from fright happens because of an affection, so too, he claims, it can happen that the body acquires this sort of coloring from a natural affection. However, in the case of fear and shame, certain *phantasmata* precede the change in coloring, while in the case of natural affections, a certain constitution does."[20] In Porphyry's view, shame and fear are both passions of the psyche, cases of psychological appearances, and bodily affections, paradigmatically changes in the coloration of the body. These passions are produced by external conditions, or in Amélie Rorty's language, by processes that are "exogenous" to the psychological process in virtue of which they come about.[21]

In order to see why Porphyry's account is convincing, one must notice Aristotle's distinctions of the several senses of *pathos* in the *Metaphysics.* Generally, *pathos* indicates either a quality in virtue of which a change occurs (e.g., the blackness or whiteness that allow for changes in color) or the actualization of that change (e.g., the hue of the body). More specifically, *pathos* denotes harmful changes and motions, as well as the big misfortunes and pains caused by such changes and motions (*Met.* 1022b15–21). These specific senses are clearly special cases of the general definitions: just as *pathos* can identify either the quality in virtue of which some change is possible or the actualized state after the change has occurred, so in the specific case, it can indicate either a potential for harmful change or the product of harmful change. In other words, *pathos* indicates both a product and a process, whether its range is general or specific. The difference is that in the specific case the kind of change must be harmful—in other words, a change in a living being that causes pain.

On the face of it, these distinctions seem to be different from the division Aristotle makes in *Categories* VIII between the four kinds of quality: dispositional states, habituated states, emotions, and the figure that gives something its shape. As Rorty has argued, in the *Categories* emotions are construed as changes in quality, specifically as changes in the qualities of the body of an animal.[22] Thus, while the *Metaphysics* division foregrounds the process-product distinction in both the general and in the animal case, the *Categories* distinction focuses on the relative permanence of the change involved. But in both texts, the divisions come to the same conclusion, which is that passions in the strict sense are short-term changes of quality in an animal.

It is unsurprising, then, that in the *Categories* Aristotle makes a further subdivision of affections into those of the body and those of the psyche, arguing that the former are not invariably caused by the latter. "It is clear that many changes in color come to be through emotions. When one is ashamed, one blushes; when afraid one becomes pale, and so on for each [of the passions]. But also if one is by nature disposed to some such affection, it arises from physiological predispositions, it might just the same be that [one has] such a complexion of skin" (*Cat.* 11b23–28). Aristotle's primary purpose here is to distinguish cases in which one's body is affected by a passion from cases in which one is disposed constitutionally to such change. For example, someone who has fair skin and little blood tends to become pale, and someone with rosacea tends to become red. In these cases, paleness or redness are not passions but dispositional bodily states, since even when they are not affected by anything external their possessors turn pale or red, and are so "by nature" (*Cat.* 12a3–9). In distinction to such dispositional changes, however, when someone turns pale from fear at the appearance of an enemy, the sight of one's protector or the realization that what appeared fearful was in fact friendly will counteract and diminish the affection.

Constitutional blushing and blushing due to a psychological passion are distinct, then, in that it is only when one is affected by a passion that the blush can be suddenly and exogenously removed. Qualities that are constitutional and long-lasting do not count as passions, such that "neither is such blushing called a blush through shame, nor such paling a paleness through fear" (*Cat.* 12a12–16). In other words, the effect of shame—the blush—is an affection in the product sense and is a *pathos* of the body rather than of the psyche. But this product must arise as the outcome of a process—namely, the psychological activity of imagining one's disrepute in the eyes of those one admires.

This coincidence of psychological emotions and bodily affections shows how passions are common to the psyche and body. When one imagines the disgrace incurred by one's actions, this imagining has an effect on the body, causing it to turn red and undergo other physiological changes. These physiological changes cause one to experience pain. The blush, in other words, is the expression of the imagined disrepute, while the pain is the effect of the physiological process marked by blushing.[23]

Indeed, Aristotle sometimes invokes this mutual action of the psyche and body on one another to distinguish episodes of a passion as such. Thus, "those who are ashamed blush, whereas those who fear death turn pale. So both appear to be somehow bodily, on account of which it seems they are passions rather than dispositional states" (EN 1128b11–16). Though the soul is the efficient cause of the bodily affection, that affection is in turn the efficient cause of the pain one experiences during the emotional episode.[24]

Though a passion could be given a purely physiological or purely dialectical description, no passion could be observed outside an ensouled body of a living animal. As the peripatetic treatise on *Physiognomics* puts it, psyche and body "are affected sympathetically by one another: on the one hand, an alteration of the state of the psyche produces an alteration in the shape of the body, and contrariwise an alteration in bodily state produces an alteration in the state of the psyche."[25] The Aristotelian account of passions is thus not perfectly analogous to the example of the form of straightness that is imprinted on metal. Rather, shame is hylomorphic in the sense that it counts both as a psychological activity of imagining one's disrepute and as a physiological change that produces pain.

This interpretation shows how psychological events that are fundamentally cognitive in character can be accompanied by sensory pleasures and pains. These are sensory because they are the result of bodily affections like blushing. Though the psyche is both the formal and the efficient cause of the bodily blush, this bodily effect is itself the efficient cause of the sensory pain experienced as part of being ashamed. This in turn clarifies the basis on which Aristotle distinguishes emotions from powers and states of the soul. While powers and states can have bodily effects, such as when one's beliefs cause one to pursue an object of desire, they need not cause one sensory pleasure or pain. Passions, in distinction, are necessarily accompanied by pleasurable or painful bodily states.

This interpretation of Aristotle's conception of shame has emphasized the sociality or intersubjectivity of shame, on the one hand, and its status as a hylomorphic state, on the other. One's immediate experience of pleasure or pain and one's awareness of one's position within a community are linked, according to this interpretation, through the middle term of imagination. By making one—incompletely and fallibly—aware of one's moral status in the eyes of others, shame links one's wishes and aspirations to the opinions of others. Yet one could imagine one's own actions from the perspective of another person and not feel shame, such as when one takes the action to be worthy of praise. However, to determine that an action is praiseworthy requires not only the judgment that it is good but also that it is fine and should be done for its own sake (EE 1248b17–20). For instance, while ends like health and strength are good, they do not necessarily contribute to the highest good, and as such are not to be chosen for their own sake. In distinction, for an action to be fine, both the wish for it and the outcome achieved must be praiseworthy.

There are two characteristics of praiseworthiness that help explain how one can revise one's ends. The first is that Aristotle thinks of truthfulness as praiseworthy, suggesting that reflection on whether one's aim is worthy of praise will make one less likely to deceive oneself or others (EN 1127a28–30). Second, there is a preferential aspect of praiseworthiness that Aristotle invokes when he describes how a good agent prefers giving money and honors to his friends to keeping them for himself. Rendering these goods to others is praiseworthy because, in so doing, one selects a good that is preferable to all the others, the fine. Consequently, even when one allows one's friend to do something fine, acting on this preference he "takes for himself" the finer, and is thus praiseworthy (EN 1169a30–b1).[26] This is also why, in both good and bad people, the reasoning part of the psyche is praised: it is this part of the psyche that summons one to do what is best (EN 1102b14–16). In other words, something is praiseworthy in virtue of its being comparable to and better than something else.

Aristotle states that choice of the fine and choice of the shameful are opposites (EN 1104b30–32). But for an action to count as praiseworthy, it must be active and go beyond the mere avoidance of shame. Episodes in which one thinks about praise and shame consist not just in thinking about the attitudes of others but also constitute an evaluation of one's own aims in terms of those attitudes. Both involve a choice of which "others" one ought to think about, and this choice determines much about the kind of life a person lives. If one chooses good friends and considers one's aims in terms of what they would find shameful or praiseworthy, one is apt to develop one's character and live a happy life. If one chooses bad friends or models, one will likely develop a distorted view of the world. Perhaps this is why Aristotle holds that even the uninformed *phantasiai* of better people tend to be about good things (EE 1219b24–25, cf. EN 1102b4–11). Merely imagining does not count as deliberation or wishing but operates on the most basic level of self-reflection, a level on which one coordinates one's internal appearances in accordance with the psyches of others.

The intimate connection between praiseworthiness and shame therefore turns out to be something other than mere opposition. Sometimes one is praised for mixed actions, when by undergoing something painful or shameful, one gains something great or fine. But one is subject to shame when one endures disgraces for something that is not fine or something that is merely mediocre (EN 1110a19–23). A consequence of this is that it is not the form of desire that makes something praiseworthy but rather the end to which that desire drives one. Thus, "appetites for the fine are also praiseworthy, those for what is shameful are blameworthy" (EN 1175b27–29). Praise and blame are defined relative to the options one has, and what makes an action worthy of shame is picking the worse end. Likewise, passively avoiding shame is a necessary but not a sufficient condition for praise, since the latter depends on an active effort to seek the fine.

The view for which I have argued may be outlined as follows. The capacity to experience shame is a criterion for living a fully human life in which one acts voluntarily, as an actor with political standing. Similarly, only voluntary actions can be praiseworthy, since they are undertaken when one has an end in view. Judgments of what is shameful and what is praiseworthy determine how one formulates and revises one's ends, and thus require the imagination, by which one can retrospectively and prospectively envision situations. Common judgments of whether an aim is shameful or praiseworthy determine norms and give a community its distinctive character. Finally, considerations of praiseworthiness and shamefulness can arise only in living beings that have the capacity to present absent situations and processes to themselves that can be used in the formulation of opinions and judgments. Aristotle has, then, taken observable social phenomena—those of people praising some individuals while blaming others, and of people designating some activities as shameful and others as fine—and has given it a psychological interpretation. According to this interpretation, these activities are not just fundamental processes of social and political life but are expressions of those inner capacities in virtue of which we are the kinds of beings we are.

EPILOGUE

> This curious property of the imagination to be set a-going by a trifle and to run on by itself in fanciful directions to extravagant lengths, and to end by forming pictures that are complete even to minute details, must be accepted as a fact for which it is not difficult simply to see a rational interpretation. Even a kaleidoscope, which consists only of two small strips of mirrors, is adequate to compose an indefinite number of tasteful and complicated patterns out of glimpses of bits of colored glass, tossed into haphazard arrangements. Much more may we suppose that the brain, whose structure is enormously complicated and acted on by organic memories as well as by present stimuli, should be capable of doing a vast deal more and something of the same kind as the very simple but suggestive instrument, the kaleidoscope.
>
> **—F. GALTON**

Ordinary attributions of cognitive phenomena to the imagination are myriad. In a dream, one asks one's father a question but wakes up to discover the conversation was imagined, not real. A geometrical proof requires a bit of imagination, when one must envisage a point moving along the circumference of a circle. Before deciding what to do, one sets a goal and imagines various courses of action by which one might achieve it. Within the ordinary attitude, these attributions are no more surprising than direct reports of perceptual experience. Imagination is part of an inherited linguistic capacity that allows one to describe and explain a wide variety of psychological events with a simple vocabulary.

This book has argued that however quotidian such utterances may be, the concept of the imagination is not something natural, like a mountain or a tree. It is the accomplishment of a psychological theory that was motivated by extraordinary concerns about our epistemic entitlement to the propositions of geometry. This entitlement could be secured neither by perception, which concerns particular material things, nor by discursive or logical reasoning, which concerns unchanging definitions. But Plato had called art *phantastikē* when it causes one to blow things out of proportion, magnify small differences, or distort things by putting lines together or taking them apart. And Plato said that there is a scene-painter in the psyche. Aristotle put the two Platonic notions together to postulate a psychological power in virtue of which we have appearances of absent and unreal things, the imagination. This power was a movement in the psyche producing and presenting "as if" motions and changes in things. This power not only furnished a coherent account of geometrical cognition but led to fruitful explanations of illusory appearances, hypnagogic experience, dreaming, retrieving memories, and making and analyzing diagrams. It explained how wish can fix on an object. One then used the imagination again, in deliberating and presenting various courses of action to reasoned judgment. The imagination explains a lot, perhaps too much.

Any historical reconstruction of the reasons for which imagination was postulated must proceed by conjecture from partial evidence. But I think it is safe to say the invention of imagination was not the solitary work of a great man. A two-hundred-year conversation about images and appearances provided the necessary ingredients for Aristotle's concoction. Plato's puzzle concerning the apparent motion of the unchanging objects of mathematics and the technique of individuating psychological powers by their objects were necessary conditions for Aristotle's employment of *phantasia* in a technical sense as a power of the psyche. Nevertheless, the data suggest that appearances that are "made" and "seen" by the psyche are of relatively recent origin.

The radical nature of Aristotle's invention is obscured by the subsequent role of *phantasia* in the philosophical tradition. Zeno the Stoic reinterprets the word to mean a presentation received by the senses, to which one could assent or not assent.[1] In the Hellenistic and Roman schools, it retains this sense as an appearance to the mind rather than as an active power. Thus, Epictetus explains there can be no good for plants, since they lack *phantasias*, impressions that present something as good or bad.[2] For the Stoic, the very essence of one's goodness consists in the proper use of these impressions.[3] A markedly psychological sense of the word is reinstituted by Augustine, who considers *phantasiae* to be *imagines* formed in the mind, by sensory impressions, by the imagination itself, or by reason.[4] Now one finds the phantasmata made and seen by a creature are something truly internal—that is, purely subjective and private impressions in the mind of an individual subject. In philosophical modernity, the imagination becomes the locus for a battle waged between the empirical materialists and the rationalistic idealists. The former follow Hobbes in conceiving of the imagination as a "decaying sense" that copies an original sensation, thus reducing higher cognition to sensory consciousness.[5] The latter follow Kant in attributing a productive and conceptual function of the imagination, by which empirical consciousness is made conformable to the "pure," nonsensory concepts of the understanding.[6] If in modern theories the imagination is the capacity to have mental images, the philosophical questions concern what counts as mental, and what counts as an image.

A parallel debate concerns whether the mental image is a legitimate concept of empirical psychological theory. Suppose mental images are generated by the intersection of two classes of neurophysiological processes: those that activate stored shapes and those that arrange these shapes into an image. And suppose that if a phenomenon identified by a psychological subject can be shown to be produced by the brain, then one should say that the phenomenon is real.[7] In that case, mental images may be real psychological simulations, which function to "generate specific predictions based on past experience."[8] Others argue that neuroscientific evidence does not support the view that mental imagery is inherently spatial and depictive but should be considered an example of a broader class of symbolic thinking. All the purported functions of mental images might be explained by spatial displays in the world, rather than images "in the head."[9] If the criterion for a phenomenon being real is its invariance across cases, mental images are not real.

The debate about mental imagery is not, of course, a question of empirical evidence. It is a question of what can be said to be "real" based on that evidence. Perhaps this puzzle of mental imagery will persist until the relationship of imagination to intellect has been radically clarified.[10] In any case, the question is bound up with our learned capacity to attribute imaginings to ourselves and others and, ipso facto, to distinguish what is merely apparent from what is real. Habits of ordinary thought and language encroach on theory. One would not know whether to ask after mental images were it not for one's preexisting conceptions of oneself and others. The capacity to be aware of and report one's experience is the very bedrock of inquiry. But this capacity is itself subject to critical historical examination. If the argument of this book is correct, such analysis reveals that the "obviousness" of the phenomena of the imagination is a sort of historical mirage, which makes invisible metaphysical and epistemological assumptions that are perhaps quite alien to us.

Consider Francis Galton's description of "the property of the imagination to be set a-going by a trifle and to run on by itself in fanciful directions to extravagant lengths" as a "fact." Only because Galton already "knew" what the imagination was could he compare the brain as an image-producing instrument to a kaleidoscope that produces a seemingly endless series of images from the juxtaposition of mirrors.

Galton's hypothetical explanation of the phenomenon of mental images depended on their already having a meaning, which could be understood both by the scientist and by his subjects. Yet upon surveying his scientific colleagues, Galton found that many had no idea what he was talking about. The men of science he surveyed denied any but a metaphorical connection between thinking and looking at the picture of an object. Galton writes the following of the scientists' attitudes toward the mental image: "They had no more notion of its true nature than a colour-blind man who has not discerned his defect has of the nature of colour."[11] Galton's survey led him to the curious conclusion that training in general and abstract thought is detrimental to the capacity to experience mental images.[12]

One is still apt to think of flexible cognition as a faculty, and to characterize is in depictive or imagistic terms. Who hasn't drawn on the "whiteboard" of one's mind, rehearsed scenes in one's "internal movie theater," or moved objects around in an inner "work space"? However imperfect they may be, these metaphors seem to express something about shared psychological experience. The free, depictive activity accompanying thinking is something one talks about with others and experiences oneself. But in examining how the imagination was invented and put to work in describing psychological phenomena, this book suggests that at least this form of self-consciousness is historically constituted and contingent. If one understands that apperception may be an "immediate" awareness of oneself, there might be experiences that are in a certain sense "infallible." The wagon may not be red, but how could one doubt that it looks red when one paints it as such in one's inner work space? Without jettisoning apperception altogether, one must note the contingency of this vocabulary of psychological concepts. Token mental states are recognizable as such only because one has learned to divide and attribute them according to types. One seems to grasp the red wagon in one's internal theater or whiteboard "immediately." But that experience is constituted by one's capacity to speak and think of oneself as subject to psychological appearances. And that capacity turns out to be not an innate idea but a "pop psychology" understanding of Aristotle's psychological theory that one has imperceptibly adopted as one's own.

The upshot of this argument is, in a way, that the imagination was itself only ever imagined. Such a general power of cognition, of mindedness, never really picked out a distinct capacity of mind. This is not just to say that the term "imagination" has a wide and indefinite extension, though this is certainly true in Aristotle's works. It's that the canonical phenomena associated with the imagination are themselves quite slippery. Imagining that the sun is the size of one's foot means what?

It's seeming that way for a moment, despite one's incredulity? A picturing of a flaming orb with a diameter equal to the length between one's heel and toes? The production of a schema that can be manipulated in various ways—for example, allowing the size of the represented objects to vary in order to make evident their topological features? But all of these mean: to represent something to oneself—that is, to think it. Is it any surprise, then, that phantasmata accompany every thought? Suppose that they are the embodied aspect of pure thinking. This negative characterization means to prevent the identification of imagining with cognition as such. But it does not make any clearer what one is doing when one is doing any of these things.

What can be said positively about the experience of imagining? One seems to entertain notions passively, to daydream, to put together incongruous elements, to blow things out of proportion. One actively envisions, pictures, and plans. But this is what one does almost all the time one is awake, and perhaps some of the time when one is asleep. To think, to wish for things to seem a certain way or appear under a particular aspect—this is the very substance of human mental life. And for this very reason, a special power or faculty that explains the quotidian substance of cognition appears to be vacuous. Just as Nietzsche's Tübingen theologians, following their master Kant, went into the bushes looking for faculties, so Aristotle, following Plato, seemed to discover a power that was altogether new. Yet Aristotle's naivete was not to mistake invention for discovery, like Nietzsche's theologians, but to imagine that an activity that was so general that it would be implicated in every act of cognition could be accounted for by a single psychological power.

Yet folks seem to take pleasure in attributing imaginings to themselves, to discussing their imaginings with others, sometimes even to comparing the imaginations of others. Imagination rarely goes out of style or is considered a topic of little interest. Artists are hailed as visionaries, prophets of the imagination. Pedagogues are concerned either with cultivating the imagination of students or in preventing them from running wild. Politicians account for poor planning in the face of uncertainty as resulting from the failure of imagination. In this sense, the imagination's value as a folk-psychological concept is very high. It plays a significant role in how people understand themselves and others. Though we cannot say exactly what it is, we recognize one another as having a sort of inner life of subjective appearances, even of images and ideas. And this recognition of something common, insofar as it produces community in understanding and action, has a real effect. One does not know what others imagine and cannot say what one is doing when one imagines something. Yet we have every reason to believe that the imagination's function in our commonsense understandings of ourselves, which emerged in the fourth century before the Christian era in Athens, is still very much alive today.

LIST *of* ABBREVIATIONS

ABBREVIATIONS OF ARISTOTLE'S WORKS (LATIN IN PARENTHESES)

AnPr.	*Prior Analytics* (*Analytica Priora*)
AnPo.	*Posterior Analytics* (*Analytica Posteriora*)
Cat.	*Categories* (*Categoriae*)
DA	*On the Soul* (*De Anima*)
Div.	*On Prophecy in Sleep* (*De Divinatione per Somnum*)
EE	*Eudemian Ethics* (*Ethics Eudemia*)
EN	*Nichomachean Ethics* (*Ethica Nicomachea*)
GA	*On the Generation of Animals* (*De Generatione Animalium*)
GC	*On Generation and Corruption* (*De Generatione et Corruptione*)
HA	*History of Animals* (*Historia Animalium*)
Insom.	*On Dreams* (*De Insomniis*)
MA	*On the Motion of Animals* (*De Motu Animalium*)
Mem.	*On Memory* (*De Memoria*)
Met.	*Metaphysics* (*Metaphysica*)
PA	*On the Parts of Animals* (*De Partibus Animalium*)
Phys.	*Physics* (*Physica*)
Sens.	*On Sense and Sense-Objects* (*De Sensu et Sensibilibus*)
Somn.	*On Sleep and Waking* (*De Somno et Vigilia*)
SophElen.	*Sophistical Refutations* (*Sophisticii Elenchii*)
Top.	*Topics* (*Topica*)

ABBREVIATIONS OF PLATO'S WORKS

Crat.	*Cratylus*
Euth.	*Euthydemus*
Gorg.	*Gorgias*
Meno	*Meno*

Parm.	*Parmenides*
Pol.	*Republic*
Phaedo	*Phaedo*
Phaed.	*Phaedrus*
Phil.	*Philebus*
Soph.	*Sophist*
Theaet.	*Thaeatetus*
Tim.	*Timaeus*

ABBREVIATION OF OTHER ANCIENT WORKS

Alex. Alexander of Aphrodisias. *On Aristotle's Metaphysics.* Edited by William E. Dooley. Ithaca, NY: Cornell University Press, 1989.

OTS Alexander of Aphrodisias. *On the Soul.* Translated by Victor Caston. London: Bloomsbury, 2012.

DK Diels, Hermann, ed. *Die Fragmente der Vorsokratiker.* Erster Band, Berlin: Weidmannsche Buchhandlung, 1906; Zweiter Band, Berlin: Weidmannsche Buchhandlung, 1922.

Elem. Euclid. *The Thirteen Books of Euclid's Elements.* Translated by Thomas L. Heath. New York: Dover, 1956.

Epist. Augustine. *Epistulae.* Edited by J. H. Baxter. Cambridge, MA: Harvard University Press, 1930.

Il. *The Iliad of Homer.* Translated by Richmond Lattimore. Chicago: University of Chicago Press, 1951.

InEucl. Proclus. *A Commentary on the First Book of Euclid's Elements.* Translated by Glenn R. Morrow. Princeton, NJ: Princeton University Press, 1970.

Philop. Philoponus. *On Aristotle's On the Soul 1.1–2.* Translated by Philip J. van der Eijk. Ithaca, NY: Cornell University Press, 2005.

Porph. *Porphyry. On Aristotle's Categories.* Translated by Steven K. Strange. Ithaca, NY: Cornell University Press, 1992.

Simplic. Simplicius. *On Aristotle's On the Soul.* Translated by H. J. Blumenthal. Ithaca, NY: Cornell University Press, 2000.

NOTES

PROLOGUE

Epigraph: Nietzsche, *Beyond Good and Evil*, §11.
1. Sellars, "Empiricism and the Philosophy of Mind," 189.
2. Sellars, *Science, Perception and Reality*, 7.
3. See Sorabji, "Concept of the Will," 8–9.
4. Jonas, *Problemi di libertà*, 255.
5. Descartes, *Principles of Philosophy*, §§34–37.
6. Lakatos, *Criticism and the Growth of Knowledge*, 132–38.
7. Dewey, *Essays in Experimental Logic*, 30.
8. Iamblichus, *De Anima*, §12.
9. Nagel, *Structure of Science*, 146–50.
10. Sellars, *Science, Perception and Reality*, 24–25.
11. Sellars, *Science, Perception and Reality*, 20.
12. Wittgenstein, *Philosophical Investigations*, §29.
13. Alexander of Aphrodisias, *On the Soul*, 1, 5 *foliis*.

CHAPTER 1: IMAGES AND APPEARANCES

1. See Jonas, *Phenomenon of Life*, 159–60.
2. Homer, *Iliad* 23.103–7.
3. Garland, *Greek Way of Death*, 12.
4. Dodds, *Greeks and the Irrational*, 104–5.
5. Vernant, *Myth and Thought*, 339–40.
6. Hobbes, *Leviathan*, 220.
7. Vernant, *Myth and Thought*, 343.
8. Vernant, *Myth and Thought*, 344–45.
9. Nicholson, *Aristocracy and Athletics*, 14–15.
10. Nicholson, *Aristocracy and Athletics*, 1, 11–12.
11. Vernant, *Myth and Thought*, 348.

12. Euripides, *Alcestis*, 348–56. Trans. Kocacs, modified.

13. Gombrich, *Art and Illusion*, 101–2.

14. Steiner, *Images in Mind*, 7–8. See Herodotus's *Histories* 4.14–15.

15. Steiner, *Images in Mind*, 9.

16. Steiner, *Images in Mind*, 10–11.

17. DK 31B 23.

18. DK 21B 2.15–19. Trans. Miller.

19. DK 21B 23. Trans. Miller.

20. DK 21B 14. Trans. Miller.

21. DK 21B 16. Trans. Miller.

22. DK 21B 15. Trans. Miller, modified.

23. DK 21 B11. Trans. Miller, modified.

24. Havelock, *Preface to Plato*, 266.

25. Havelock, *Preface to Plato*, 258–59.

26. Havelock, *Preface to Plato*, 261.

27. Harrison, *Ancient Art and Ritual*, 21–23.

28. DK 68A 135.

29. Aetius IV, 8, 10. For discussion of these passages, see Kirk and Raven, *Presocratic Philosophers*, 421–24.

30. DK 80B 1. Trans. Freeman.

31. Sextus Empiricus, *Outlines of Pyrrhomism*, 216–17.

32. de Romilly, *Great Sophists*, 189–90.

33. de Romilly, *Great Sophists*, 191–92.

34. de Romilly, *Great Sophists*, 196–212.

35. de Romilly, *Great Sophists*, 97–103.

36. McDowell, *Plato Theaetetus*, 369.

37. This is just one way of understanding Plato's argument. For a more systematic discussion of the reasons Plato might have had for taking this argument to be self-refuting, see Keeling, "Truth for a Person and Truth for a *Polis*."

38. For discussion of earlier movements toward formalization, see Knorr, *Ancient Tradition of Geometric Problems*, 324–25.

39. Herodotus, *Histories* 1.108–9.

40. See Klein, *Commentary on Plato's Meno*, 64–67 for discussion. Klein suggests that the naive account is better in some respects.

41. See discussion in Klein, *Greek Mathematical Thought*, 22–24.

42. Tait, "Noēsis," 4–6.

43. Cornford, "Mathematics and Dialectic in the Republic (I.)," 41–42.

44. Cornford, "Mathematics and Dialectic in the Republic (I.)," 39–40.

45. See Proclus, *InEucl.* 76. Outside of the *Republic*, the meaning of hypothesis is less certain (e.g. at *Phaedo* 100a).

46. Reeve, *Philosopher-Kings*, 54–55.

47. Reeve, *Philosopher-Kings*, 287n9.

48. Cornford, "Mathematics and Dialectic in the Republic (I.)," 38f.

49. Hackforth, "Plato's Divided Line and Dialectic," 1–2.

50. Hackforth, "Plato's Divided Line and Dialectic," 8.

51. Burnyeat, "Plato on Why Mathematics is Good for the Soul," 22.

52. Cross and Woozley, *Plato's Republic*, 234.

53. Aristotle makes a similar argument on behalf of the Platonist (*Met.* 1002b12–24), in which he suggests that only a doctrine of intermediates can account for why there are many mathematical objects of the same kind.

54. Cross and Woozley, *Plato's Republic*, 236–237.

55. Proclus *In Tim* I.8.

56. Ivor Bulmer-Thomas argues that Plato's methodological comments on astronomy are intended to shift astronomy from grouping "stars into fanciful figures" to focus on building "a mathematical explanation of celestial motions" (Bulmer-Thomas, "Plato's Astronomy," 112).

57. Dillon, *Heirs of Plato*, 43–44.

58. Dillon, *Heirs of Plato*, 47–50.

CHAPTER 2: POWERS OF THE PSYCHE

1. Cherniss, "Plato as Mathematician," 423–24.

2. Proclus also connects Aristotle and Amphinomous *In Eucl.* 253–54. See Knorr, *Ancient Tradition of Geometric Problems*, 74–75 for discussion of Amphinomus's possible influence on Aristotle's *Posterior Analytics* An Po. I.4.12.

3. Lachterman, *Ethics of Geometry*, 63. Trans. modified.

4. Lachterman, *Ethics of Geometry*, 92, cf. 94 for discussion of the "homogeneity" of the components of geometrical figures.

5. See Knorr, *Ancient Tradition of Geometric Problems*, 60–62 for discussion of Menaechmus's technique of cube duplication.

6. Indeed, it is unclear whether the scheme correctly describes pure construction problems. See Netz, *Shaping of Deduction in Greek Mathematics*, 252–63 for discussion.

7. Cf. *Philebus* 20a.

8. Jaeger, *Aristotle*, 19–20.

9. Dillon, *Heirs of Plato*, 27–29, 79–82, 115–17.

10. For a lucid exposition of the relevant passages in Plato and Aristotle, see Johansen, *Powers of Aristotle's Soul*, 94–95.

11. Johansen, *Powers of Aristotle's Soul*, 1, 56–57.

12. Lear, *Aristotle*, 117.

13. Weinman, *Pleasure in Aristotle's Ethics*, 6.

14. Silverman, "Plato on Phantasia," 124, 132–33.

15. See Schofield, "Aristotle on the Imagination," 99; cf. Dillon, *Alcinous*, 7, 17.

16. O'Connor, "Rewriting the Poets in Plato's Characters," 81–82.

17. See Cornford, *Plato's Theory of Knowledge*, 32n3.

CHAPTER 3: IMAGINATION DEFINED

1. This is because conviction is propositional—that is, it consists in being committed to a λόγος. Whether this conception of conviction agrees or disagrees with Plato's placement of it in the divided line is an open question. It speaks against the combinationist position represented in the *Sophist*, since that view would limit phantasia to animals with language (DA 428a24–29).

2. For discussion of this sort of productive art, see *Phys*. 194a34–b8.

3. This presentational function has been noticed by Yurdin, "Aristotelian Imagination and the Explanation of Behavior."

4. Schofield, "Aristotle on the Imagination," 106. Cf. Moss, *Aristotle on the Apparent Good*, xii.

5. Nussbaum, *Aristotle's De Motu Animalium*, 230–31.

6. Nussbaum, *Aristotle's De Motu Animalium*, 260–61.

7. Sorabji, *Aristotle on Memory*, 72, cf. xiv–xx for further discussion.

8. Diamond, *Mortal Imitations of Divine Life*, 160.

9. Scheiter, "Images, Appearances, and Phantasia in Aristotle," 259–60.

10. Johansen, *Powers of Aristotle's Soul*, 203.

11. Several modern readers claim that Aristotle considers the imagination solely to reproduce the contents of perception. See, for example, Schofield, "Aristotle on the Imagination," 114; Frede, "Cognitive Role of Phantasia in Aristotle," 282, cf. 289; Moss, *Aristotle on the Apparent Good*, 51–53; and Johansen, *Powers of Aristotle's Soul*, 202.

CHAPTER 4: DREAMING AND DAY-DREAMING

1. DK 31B 105, trans. Freeman.

2. Burnyeat, "'De Anima' I.5," 44–45.

3. Ross, *Aristotle*, 130.

4. Cf. DA 428a9–11, where Aristotle attributes the imagination to ants and bees. There is a tension, if not an outright contradiction, between the passages, however, since one seems to say that all animals participate in imagination, whereas the other excludes the grub, and perhaps other animals that live by touch alone.

5. For the example, see *De Caelo* 272b 25–29.

6. Frede, "Cognitive Role of Phantasia in Aristotle," 282.

7. Caston, "Why Aristotle Needs Imagination," 47.

8. Bolton, "Perception Naturalized in Aristotle's *De Anima*," 232.

9. See *De Anima* 428a2 for the distinction between psychological and extended or "metaphorical" attributions of phantasia. Such attributions occur (e.g., at *Topics* 100b27) to describe arguments that appear to be generally accepted, and in his discussion of "apparent light" in *De Caelo*. Even in psychological works, "appearance" words may have a general sense that does not refer to the imagination or its functions. See, for example, *De Anima* 420b31–33, 425b25; *De Sensu* 437a25–26.

10. Schofield, "Aristotle on the Imagination," 130.

11. Frede, "Cognitive Role of Phantasia in Aristotle," 284.

12. Frede, "Cognitive Role of Phantasia in Aristotle," 296.

13. Frede, "Cognitive Role of Phantasia in Aristotle," 294.

14. Moss, *Aristotle on the Apparent Good*, 63.

15. Moss, *Aristotle on the Apparent Good*, 3.

16. Moss, *Aristotle on the Apparent Good*, 11–13.

17. Polansky, *Aristotle's* De Anima, 412; cf. Lycos, "Aristotle and Plato on 'Appearing,'" 497.

18. Polansky, *Aristotle's* De Anima, 410.

19. Benardete, "Aristotle 'De Anima' III. 3–5," 614.

20. DK 22 B3.

21. While the falsity of a proposition or a state of the psyche with a propositional content such as a belief consists in a combination of things that cannot or do not go together (*Met.* 1024b17–20, 26–28), the falsity of dreams does not depend on this kind of defective composition. For although dreams are entities, the phantasia from them is not a being (*Met.* 1024b25). The example of the man walking is from Alexander (*Met.* V.432,20).

22. Ross, *Aristotle*, 143.

23. Bolton, "Perception Naturalized in Aristotle's *De Anima*," 232–33.

24. Bolton, "Perception Naturalized in Aristotle's *De Anima*," 235.

25. Ross, *Aristotle*, 144.

26. Moss, *Aristotle on the Apparent Good*, 52–54.

27. Johansen, *Powers of Aristotle's Soul*, 200–204.

28. Johansen, *Powers of Aristotle's Soul*, 203.

29. See Polansky and Fritz, "Aristotle on Accidental Perception," for defense of the claim that, for example, when one sees a triangle, one accidentally perceives an intelligible triangle.

CHAPTER 5: ABSTRACTING

1. Mueller, "Aristotle on Geometrical Objects."

2. Lear, "Aristotle's Philosophy of Mathematics."

3. Netz, "Aristotle's Metaphysics M3."

4. Corkum, "Aristotle on Mathematical Truth."

5. Unguru, "On the Need to Rewrite the History of Greek Mathematics," 69–75; cf. Unguru, "History of Ancient Mathematics," 564.

6. Unguru, "On the Need to Rewrite the History of Greek Mathematics," 76.

7. Unguru's work can be understood as an elaboration of projects begun in the 1930s by Edmund Husserl and Jakob Klein. See Husserl, *Crisis of the European Sciences*, 44–45, 353–78; Klein, *Greek Mathematical Thought*, 120–23; Unguru, "On the Need to Rewrite the History of Greek Mathematics," 109.

8. Einarson, "On Certain Terms in Aristotle's Logic," 36.

9. Pappus *Collectio* VII, 1. Trans. Jones, *Pappus of Alexandria*, 82.

10. Pappus *Collectio* VII, 1. Trans. Jones, *Pappus of Alexandria*, 82–84.

11. See Gulley, "Greek Geometrical Analysis" for a summary of the classical debate.

12. Berggren and Van Brummelen, "Role and Development of Geometrical Analysis," 9–10.

13. Barnes, *Aristotle Posterior Analytics*, 19. Trans. modified.

14. Menn, "Plato and the Method of Analysis," 194.

15. Menn, "Plato and the Method of Analysis," 205–9.

16. Knorr, *Ancient Tradition of Geometric Problems*, 9; Proclus *In Eucl.* 255.

17. Manders, "Euclidean Diagram," 111

18. Manders, "Euclidean Diagram," 92.

19. Manders, "Euclidean Diagram," 93–94.

20. Heath, *Thirteen Books of Euclid's Elements*, 116–17, 252–54.

21. See *Phaedo* 95e2–6, *Euth.* 296b, *Crat.* 393d, 432a, *Parm.* 131d, 158c.

22. Cleary, "On the Terminology of 'Abstraction' in Aristotle," 18.

23. Mueller, "Aristotle on Geometrical Objects," 160.

24. Locke, *Essay Concerning Human Understanding*, III.3.6.

25. Mueller, "Aristotle on Geometrical Objects," 170; Lear, "Aristotle's Philosophy of Mathematics," 184–86.

26. Lear, "Aristotle's Philosophy of Mathematics," 186.

27. Lear, "Aristotle's Philosophy of Mathematics," 169.

28. Lear, "Aristotle's Philosophy of Mathematics," 173.

29. This is equivalent to *Elem.* I.32, which shows that the method of "drawing beyond the side" was known quite early (Heiberg, *Mathematisches zu Arisoteles*, 19–20; cf. Heath, *Thirteen Books of Euclid's Elements*, 319–20). Eudemus the Peripatetic credits the Pythagoreans with the discovery of the proposition, giving a version of their proof in which a line parallel to BC is drawn through A (*InEucl.* 379). But Geminus attributes the discovery to Thales, implying that early proofs treated the scalene, isosceles, and

equilateral triangles as separate cases (Heath, *History of Greek Mathematics*, 135–36). Heath notes that Geminus's assertion may have arisen from a misattribution of historical meaning to Aristotle's complaint that separating the cases does not result in a universal proof. In any case, Aristotle uses the proposition as a stock example, suggesting it was well known to his audience.

30. Euclid gives the demonstration as follows: Since AB is parallel to CE, and AC has fallen upon them, the alternate angles BAC and ACE are equal to each other. Again since AB is parallel to CE, and the straight line BD has fallen upon them, the exterior angle ECD is equal to the interior and opposite angle ABC. But the angle ACE was also proved equal to the angle BAC; therefore the whole angle ACD is equal to the two interior and opposite angles BAC and ACE. Let the angle ACB be added to each. Therefore the angles ACD and ACB are equal to the three angles ABC, BCA, and CAB. But the angles ACD and ACB are equal to two right angles; therefore, the angles ABC, BCA, and CAB are also equal to two right angles (Heath, *Thirteen Books of Euclid's Elements*, 317).

CHAPTER 6: GEOMETRICAL THINKING

1. For discussion, see Katz, "Absurd Accumulation."
2. Makin, *Metaphysics, Book Theta*, 232–37.
3. Heath, *Thirteen Books of Euclid's Elements*, 63–64.
4. This assumption is proved in proposition III.21 of Euclid's *Elements*, but Euclid's III.31 does not make use of it.
5. Heath, *Thirteen Books of Euclid's Elements*, 63.
6. See *Phys.* 195a15–18, *Met.* 1013a15, 1013b17.
7. Barnes, *Aristotle Posterior Analytics*, 227.
8. Makin, *Metaphysics*, 238–40.

CHAPTER 7: WISHING

1. Boeri, "Presence of Socrates," 389–92 identifies this view with the "monist psychology" of the Stoics.
2. EN 1113a25–29, cf. *Theaet.* 152a.
3. See Nussbaum, *Fragility of Goodness*, 276.
4. For a related neo-Aristotelian discussion of form and desire, see Boyle and Lavin, "Goodness and Desire," 179.
5. Irwin, "Virtue, Praise and Success," 331–32.
6. Moss, *Aristotle on the Apparent Good*, 6–8.
7. Hume, *Treatise of Human Nature*, 458.
8. Kant, *Critique* II, 32.

9. See, for example, Hamlyn, *Aristotle De Anima*, 130–31.

10. I owe this example to Ronald Polansky.

11. Lännström, *Loving the Fine*, 11.

12. Lännström, *Loving the Fine*, 20.

13. Although a description of acrasia is beyond the scope of my discussion here, this is consistent with Aristotle's view that an acratic agent, when overcome by passion, can act against his own wishes (EN 1166b6).

CHAPTER 8: DELIBERATING

1. Wiggins, "Deliberation and Practical Reason," 223–24. See also Broadie, *Ethics with Aristotle*, 239; McDowell, "Some Issues in Aristotle's Moral Psychology," 25; and Segvic, *From Protagoras to Aristotle*, 172–73.

2. Wiggins, "Deliberation and Practical Reason," 227.

3. Modrak, "Aisthēsis in the Practical Syllogism," 383.

4. McDowell, "Virtue and Reason," 127–29; "Some Issues in Aristotle's Moral Psychology," 26–28.

5. Modrak, "Aisthēsis in the Practical Syllogism," 390n22.

6. Einarson, "On Certain Terms in Aristotle's Logic," 36.

7. Segvic, *From Protagoras to Aristotle*, 170.

8. See *Metaphysics* Z.7, 1032b6–10 for an example of this kind of reasoning in technical deliberation.

9. I here take τὰ ζῷα (DA 433b30) as the antecedent to which τῶν ἀτελῶν (DA 433b21) refers. Yet ἄλλων is a plausible alternative antecedent. In that case, Aristotle refers to "other animals" rather than to animals without ends. But this does not change the substance of my interpretation, since the "other animals" (at, e.g., DA 434a6) seem to be those for whom calculation is impossible. On either reading, Aristotle distinguishes between ensouled creatures able to make choices and those that a live only by immediate appearances, for which imagination and movement are alike indeterminate.

10. This objection was put to me by Ronald Polansky, in private correspondence.

11. Jonas, *Phenomenon of Life*, 161.

12. Lorenz, *Brute Within*, 124f is surely right to insist that perceptual and deliberative imagination are not mutually exclusive: while aesthetic imagination alone is present in animals, both are in play in humans.

13. See, for example, Hamlyn, *Aristotle De Anima*, 153.

14. Anscombe, "Thought and Action in Aristotle," 72–74.

15. To my knowledge, this important contribution to the literature on Aristotelian deliberation remains unpublished. I thank Gisela Striker for discussion on an early draft of this chapter.

16. Aristotle makes a similar point about deliberation in the *Rhetoric* I.2,

when he characterizes deliberative rhetoric as concerning things for which we do not possess technical knowledge, and that admit of being otherwise.

17. Menn, "Aim and the Argument of Aristotle's Metaphysics," chap. IIIa2, 6–7, points out that the assertion that rational powers are simultaneously for contraries is an "Academic commonplace" that goes back to Plato and to Socrates's characterization of science. According to Menn, Aristotle explains a power's production of contradictories in terms of that power becoming conjoined with other powers: thus he must exclude natures from his conception of power, presumably because natures must always bring about the same thing.

18. Sorabji, "Aristotle on the Rôle of Intellect in Virtue," 117.

CHAPTER 9: SHAME AND THE OTHER

1. Many scholars translate πάθη as "emotions" or "affects," perhaps because these are more familiar terms than "passions" in contemporary moral psychology. However, I use "passion" to refer to πάθη of the psyche, reserving "affect" to describe πάθη of the body. Both terms have the advantage over "emotion" in that they retain the passive sense of something that is suffered by an agent.

2. There are two kinds of ignorance at play in Aristotle's account. An action done in ignorance of the universal is nonvoluntary and is typically motivated by an irrational affection such as drink or wrath. Such actions may be blameworthy, when the agent is to blame for his recklessness. An action done through ignorance of a particular, in distinction, is typically not blameworthy, since such an action does not involve general ignorance and may not be caused by bad desires. These latter, blameless actions, typically involve regret.

3. Homer, *Iliad*, II.262

4. Grimaldi, *Aristotle, Rhetoric II*, 107.

5. Konstan, "Shame in Ancient Greece," 1042.

6. Ron Polansky has pointed out to me in conversation that apparent inconsistencies between Aristotle's works may often be resolved through the observation that different inquiries demand different levels of precision (see, e.g., EN 1094b22–27, 1104a1–5). Thus, in psychological and biological contexts, Aristotle grants motivating status to desires, since unless an animal is moved by force, the object of desire must be the mover (DA 432b15–17, 433a23–28, MA 700b24–29). However, in a practical, public context one can treat the sense of shame as formative of the goals an agent pursues.

7. See Berns, "Rational Animal-Political Animal."

8. Nussbaum, *Therapy of Desire*, 80–82.

9. Sherman, *Making a Necessity of Virtue*, 39, 55–58

10. Konstan, "Shame in Ancient Greece," 1042.

11. Grimaldi, *Aristotle, Rhetoric II*, 115.

12. The *Rhetoric* portrays the decision of whose opinions matter as dependent on social criteria such as status and rank, whether other agents are friends or rivals, and relations of kinship. Foremost among these relations is that of familiarity: one feels shame for what seems to be truly shameful among those one knows, while in front of those one does not, one feels shame about what is customary (*Rhet.* 1384b24–27).

13. Leighton, "Aristotle on the Emotions," 149.

14. Descartes defines emotions as thoughts of the soul caused by a movement of animal spirits in the body in *Passions of the Soul*, Part 1, Article 27. See Haldane and Ross, *Philosophical Works of Descartes*, 344.

15. Leighton, "Aristotle on the Emotions," 158, 168.

16. Fortenbaugh, "Aristotle and Theophrastus on the Emotions," 41.

17. Fortenbaugh, "Aristotle and Theophrastus on the Emotions," 7–38.

18. Aristotle holds that passions are common to body and soul at *De sensu et sensibilibus* 436b3–6. See Mansion, "Soul and Life in *De Anima*," 7 for a discussion of the methodological issues raised by the distinction between dialectical and physiological descriptions.

19. Aristotle summarizes his point about separability by stating that "the affections of the psyche, insofar at least as they are things like anger and fear, are in this way inseparable from the natural matter of living things, rather than inseparable only in the way that a line or surface are" (DA 403b15–19).

20. Porphyry in *Cat* 130, 32–131, 2. S. K. Strange, Trans.

21. Rorty, "Aristotle on the Metaphysical Status of 'Pathe,'" 524.

22. Rorty, "Aristotle on the Metaphysical Status of 'Pathe,'" 525–26.

23. Modern theories of affect have reached similar conclusions about the expression of shame. Heller, "Power of Shame," 215 draws on Darwin's work on emotion to argue that expressions of affect involve the activation of an acquired capacity and are communicative. Darwin, *Expression of the Emotions*, 310–12, however, holds that the capacity to blush is inherited and is triggered involuntarily. Moreover, Darwin, *Expression of the Emotions*, 328–33 conjectures that expressions of shame are primarily facial expressions: turning one's face away from the gaze of others or lowering the eyes, both of which are noted by Aristotle (*Rhet.* 1384a34–b2, 1385a9–13).

24. The relationship between these two events is causal in an efficient sense: one's imagining of the disrepute is what makes one blush. Thus, passions could be prior to the substance in which they inhere neither in account, nor in time, nor in genesis (*Met.* 1038b27–29).

25. *Physiognomics* (808b11–15). Hett, Trans. modified.

26. This description of the preferential aspect of praise has roots in the Platonic criticism of portraying the gods in human attitudes. Just as it

would not be praiseworthy if the gods had appetites, they must likewise have nothing to do with action, which would not be praiseworthy in a god (EN 1178b15–19). For an activity to be praiseworthy, it must be preferable to another activity, and thus for the gods to be praiseworthy, their activities must be better than those of humans.

EPILOGUE

Epigraph: Galton, "Free-Will," 411–12.

1. Cicero, *Academica*, §40.
2. Epictetus, *Discourses* II.8. 4–6.
3. "ἡ οὐσία τοῦ ἀγαθοῦ ἔστιν ἐν χρήσει φαντασιῶν καὶ τοῦ κακοῦ ὡσαύτος" (Epictetus, *Discourses* II.1, 4).
4. Augustine, *Letter VII*, §§4–5.
5. Hobbes, *Leviathan*, I.2, 4–5.
6. Kant, *Kritik der Reinen Vernunft*, A.138/B.177—A.140/B.179.
7. Kosslyn, "Aspects of a Cognitive Neuroscience of Mental Imagery."
8. Moulton and Kosslyn, "Imagining Predictions," 1274.
9. Pylyshyn, "Return of the Mental Image," 113–17.
10. Brann, "Mental Imagery," 77–78.
11. Galton, "Statistics of Mental Imagery," 302.
12. See Burbridge, "Galton's 100," for discussion of Galton's survey.

BIBLIOGRAPHY

Ackrill, J. L. "Aristotle's Distinction between *Energeia* and *Kinesis*." In *New Essays on Plato and Aristotle*, edited by Renford Bambrough, 121–41. London: Routledge, 1965.

Anscombe, G. E. M. "Thought and Action in Aristotle." In *From Parmenides to Wittgenstein*, 66–77. Minneapolis: University of Minnesota Press, 1981.

Barnes, Jonathan. *Aristotle: Posterior Analytics*. 2nd ed. Oxford: Clarendon, 2002.

Barnstone, Willis. *Sappho*. New York: Anchor Books, 1965.

Benardete, Seth. "Aristotle 'De Anima' III. 3–5." *Review of Metaphysics* 28, no. 4 (1975): 611–22.

Berggren, J. L., and Glen Van Brummelen. "The Role and Development of Geometrical Analysis and Synthesis in Ancient Greece and Medieval Islam." In *Ancient and Medieval Traditions in the Exact Sciences: Essays in Memory of Wilbur Knorr*, edited by Patrick Suppes, Julius M. Moravcsik, and Henry Mendell, 1–31. Stanford, CA: CLSI Publications, 2000.

Berns, Laurence. "Rational Animal-Political Animal: Nature and Convention in Human Speech and Politics." *Review of Politics* 38, no. 2 (1976): 177–89.

Boeri, Marcelo D. "The Presence of Socrates and Aristotle in the Stoic Account of *Akrasia*." In *Metaphysics, Soul, and Ethics in Ancient Thought: Themes from the Work of Richard Sorabji*, edited by Ricardo Salles, 382–412. Oxford: Clarendon, 2005.

Bolton, Robert. "Perception Naturalized in Aristotle's *De Anima*." In *Metaphysics, Soul, and Ethics in Ancient Thought: Themes from the Work of Richard Sorabji*, edited by Ricardo Salles, 209–44. Oxford: Oxford University Press, 2005.

Boyle, Matthew, and Douglas Lavin. "Goodness and Desire." In *Desire, Practical Reason, and the Good*, edited by Sergio Tenenbaum, 161–201. Oxford: Oxford University Press, 2010.

Brann, Eva. "Mental Imagery." *St. John's Review*, Summer 1985, 71–79.

Broadie, Sarah. *Ethics with Aristotle*. New York: Oxford University Press, 1991.

Bulmer-Thomas, Ivor. "Plato's Astronomy." *Classical Quarterly* 34, no. 1 (1984): 107–12.

Burbridge, David. "Galton's 100: An Exploration of Francis Galton's Imagery Studies." *British Journal for the History of Science* 27, no. 4 (1994): 443–63.

Burnyeat, Myles F. "'De Anima' I. 5." *Phronesis* 47, no. 1 (2002): 28–90.

Burnyeat, Myles F. "Plato on Why Mathematics Is Good for the Soul." *Proceedings of the British Academy* 103 (2000): 1–81.

Caston, Victor. "Why Aristotle Needs Imagination." *Phronesis* 41, no. 1 (1996): 20–55.

Cherniss, Harold. "Plato as Mathematician." *Review of Metaphysics* 4, no. 3 (1951): 395–425.

Cleary, John J. "On the Terminology of 'Abstraction' in Aristotle." *Phronesis* 30, no. 1 (1985): 13–45.

Corkum, Phil. "Aristotle on Mathematical Truth." *British Journal for the History of Philosophy* 20, no. 6 (2012): 1057–76.

Cornford, Francis M. "Mathematics and Dialectic in the *Republic* VI.–VII. (I.)." *Mind* 41, no. 161 (1932): 37–52.

Cornford, Francis M. "Mathematics and Dialectic in the *Republic* VI.–VII. (II.)." *Mind* 41, no. 162 (1932): 173–90.

Cornford, Francis MacDonald. *Plato's Theory of Knowledge*. New York: Liberal Arts Press, 1957.

Cross, R. C., and A. D. Woozley. *Plato's Republic: A Philosophical Commentary*. London: Macmillan, 1964.

Darwin, Charles. *The Expression of the Emotions in Man and Animals*. New York: D. Appleton and Company, 1873.

de Romilly, Jacqueline. *The Great Sophists in Periclean Athens*. Translated by Janet Lloyd. Oxford: Clarendon, 1992.

Descartes, René. *The Principles of Philosophy*. In *The Philosophical Works of Descartes*, vol. 1, translated by E. S. Haldane and G. R. T. Ross. Cambridge: Cambridge University Press, 1931.

Dewey, John. *Essays in Experimental Logic*. Chicago: University of Chicago Press, 1916.

Diamond, Eli. *Mortal Imitations of Divine Life*. Evanston, IL: Northwestern University Press, 2015.

Dillon, John. *Alcinous: The Handbook of Platonism*. Oxford: Clarendon, 1993.

Dillon, John. *The Heirs of Plato*. Oxford: Clarendon, 2003.

Dodds, Eric R. *The Greeks and the Irrational*. Berkeley: University of California Press, 1951.

Einarson, Benedict. "On Certain Terms in Aristotle's Logic: Part I." *American Journal of Philology* 57, no. 1 (1936): 33–54.

Finnamore, John F., and John M. Dillon, eds. *Iamblichus' De Anima: Text, Translation, and Commentary*. Leiden, Netherlands: Brill, 2002.

Fortenbaugh, William W. "Aristotle and Theophrastus on the Emotions." In *Passions and Moral Progress in Greco-Roman Thought*, edited by John T. Fitzgerald, 29–47. New York: Routledge, 2007.

Frede, D. "The Cognitive Role of Phantasia in Aristotle." In *Essays on Aristotle's De Anima*, edited by Martha C. Nussbaum and Amélie Oksenberg Rorty, 280–96. Oxford: Oxford University Press, 1995.

Freeman, Kathleen. *Ancilla to the Pre-Socratic Philosophers*. Cambridge, MA: Harvard University Press, 1948.

Fussi, Alessandra. "Aristotle on Shame." *Ancient Philosophy* 35, no. 1 (2015): 113–35.

Galton, Francis. "Free-Will—Observations and Inferences." *Mind* 9, no. 35 (1884): 406–13.

Galton, Francis. "Statistics of Mental Imagery." *Mind* 5, no. 19 (1880): 301–18.

Garland, Robert. *The Greek Way of Death*. Ithaca, NY: Cornell University Press, 1985.

Gombrich, E. H. *Art and Illusion: A Study in the Psychology of Pictorial Representation*. London: Phaidon, 1984.

Grimaldi, William M. A. *Aristotle, Rhetoric II: A Commentary*. New York: Fordham University Press, 1988.

Gulley, Norman. "Greek Geometrical Analysis." *Phronesis* 3, no. 1 (1958): 1–14.

Hackforth, Reginald. "Plato's Divided Line and Dialectic." *Classical Quarterly* 36, no. 1/2 (1942): 1–9.

Haldane, Elizabeth S., and G. R. T. Ross. *The Philosophical Works of Descartes*, vol. 1. Cambridge: Cambridge University Press, 1931.

Hamlyn, D. W. *Aristotle: De Anima*. Oxford: Oxford University Press, 2002.

Harrison, Jane Ellen. *Ancient Art and Ritual*. New York: Henry Holt, 1913.

Havelock, Eric A. *Preface to Plato*. Oxford: Blackwell, 1963.

Heath, Thomas. *A History of Greek Mathematics*. Vol. 1, *From Thales to Euclid*. New York: Dover, 1981.

Heiberg, J. L. *Mathematisches zu Aristoteles*. Leipzig: B. G. Teubner, 1904.

Heller, Agnes. "The Power of Shame." *Dialectical Anthropology* 6, no. 3 (1982): 215–28.

Herodotus. *The Histories*. Edited by Aubrey de Sélincourt. Harmondsworth, UK: Penguin, 1954.

Hobbes, Thomas. *Leviathan*. London: Penguin Classics, 1985.

Hume, David. *A Treatise of Human Nature*. Edited by L. A. Selby-Bigge. Oxford: Clarendon, 1846.

Husserl, Edmund. *The Crisis of the European Sciences and Transcendental Phenomenology*. Translated by David Carr. Evanston, IL: Northwestern University Press, 1970.

Irwin, Terence H. "Aristotle on Reason, Desire, and Virtue." *Journal of Philosophy* 72, no. 176 (1975): 567–68.

Irwin, Terence H. *Aristotle's First Principles*. Oxford: Clarendon, 1988.

Irwin, Terence H. "Virtue, Praise and Success: Stoic Responses to Aristotle." *Monist* 73, no. 1 (1990): 59–79.

Jaeger, Werner. *Aristotle: Fundamentals of the History of His Development*. 2nd ed. Translated by Richard Robinson. Oxford: Clarendon, 1960.

Johansen, Thomas Kjeller. *The Powers of Aristotle's Soul*. Oxford: Oxford University Press, 2012.

Jonas, Hans. *The Phenomenon of Life*. Evanston, IL: Northwestern University Press, 2001.

Jonas, Hans. *Problemi di libertà*. Turin: Nino Aragno Editore, 2010.

Jones, Alexander, ed. *Pappus of Alexandria Book 7 of the Collection. Part 1. Introduction, Text, and Translation*. New York: Springer, 1986.

Kant, Immanuel. *Critique of Practical Reason*. Translated by T. K. Abbot. Amherst, NY: Prometheus Books, 1996.

Katz, Emily. "An Absurd Accumulation: *Metaphysics* M. 2, 1076b11–36." *Phronesis* 59 (2014): 343–68.

Keeling, Evan. "Truth for a Person and Truth for a *Polis*: A Note on *Theaetetus* 171a1–6." *Ancient Philosophy* 35, no. 1 (2015): 63–73.

Kirk, G. S., and J. E. Raven. *The Presocratic Philosophers*. Cambridge: Cambridge University Press, 1966.

Klein, Jacob. *A Commentary on Plato's Meno*. Chapel Hill: University of North Carolina Press, 1965.

Klein, J. *Greek Mathematical Thought and the Origin of Algebra*. Translated by Eva Brann. Cambridge, MA: MIT Press, 1968.

Knorr, Wilbur R. *The Ancient Tradition of Geometric Problems*. New York: Dover, 1986.

Knorr, Wilbur R. "The Practical Element in Ancient Exact Sciences." *Synthese* 81, no. 3 (1989): 313–28.

Konstan, David. "Shame in Ancient Greece." *Social Research* 70, no. 4 (2003): 1031–60.

Kosslyn, Stephen M. "Aspects of a Cognitive Neuroscience of Mental Imagery." *Science*, n.s. 240, no. 4859 (1988): 1621–26.

Kozhevnikov, Maria, Stephen Kosslyn, and Jennifer Shephard. "Spatial versus Object Visualizers: A New Characterization of Visual Cognitive Style." *Memory and Cognition* 33, no. 4 (2005): 710–26.

Lachterman, David Rapport. *The Ethics of Geometry*. New York: Routledge, 1989.

Lakatos, Imre. *Criticism and the Growth of Knowledge*. Cambridge: Cambridge University Press, 1970.

Lännström, Anna. *Loving the Fine: Virtue and Happiness in Aristotle's Ethics*. Notre Dame, IN: Notre Dame University Press, 2006.

Lear, Jonathan. *Aristotle: The Desire to Understand*. Cambridge: Cambridge University Press, 1988.

Lear, Jonathan. "Aristotle's Philosophy of Mathematics." *Philosophical Review* 91, no. 2 (1982): 161–92.

Leighton, Stephen R. "Aristotle on the Emotions." *Phronesis* 27, no. 2 (1982): 144–74.

Locke, John. *An Essay Concerning Human Understanding*. Edited by Pauline Phemister. Oxford: Oxford University Press, 2008.

Lorenz, Hendrik. *The Brute Within: Appetitive Desire in Plato and Aristotle*. Oxford: Oxford University Press, 2006.

Löwith, Karl. *From Hegel to Nietzsche: The Revolution in Nineteenth-Century Thought*. New York: Columbia University Press, 1964.

Lycos, Kimon. "Aristotle and Plato on 'Appearing.'" *Mind*, n.s. 73, no. 292 (1964): 496–514.

Makin, Stephen. *Aristotle: Metaphysics, Book Theta*. Oxford: Oxford University Press, 2006.

Malcolm, Norman. "Dreaming and Skepticism." In *Descartes: A Collection of Critical Essays*, edited by Willis. Doney, 54–79. New York: Anchor Books, 1967.

Manders, Kenneth. "The Euclidean Diagram." In *The Philosophy of Mathematical Practice*, edited by Paolo Mancosu, 80–133. Oxford: Oxford University Press, 2008 [1995].

Mansion, S. "Soul and Life in *De Anima*." In *Aristotle on Mind and the Senses: Proceedings of the Seventh Symposium Aristotelicum*, edited by G. E. R. Lloyd and G. E. L. Owen, 1–20. London: Cambridge University Press, 1978.

McDowell, John. "Deliberation and Moral Development in Aristotle's Ethics." In *Aristotle, Kant, and the Stoics: Rethinking Happiness and Duty*, edited by Jennifer Whiting and Stephen Engstrom, 19–35. Cambridge: Cambridge University Press, 1998.

McDowell, John, trans. *Plato: Theaetetus*. Oxford: Clarendon, 1973.

McDowell, John. "Some Issues in Aristotle's Moral Psychology." In *Mind, Value, and Reality*, 23–49. Cambridge, MA: Harvard University Press, 1998.

McDowell, John. "Virtue and Reason." In *Aristotle's Ethics: Critical Essays*, edited by Nancy Sherman, 121–43. New York: Rowman and Littlefield, 1999.

Menn, Stephen. "The Aim and the Argument of Aristotle's Metaphysics." Unpublished manuscript, 2015. https://www.philosophie.hu-berlin.de/de/lehrbereiche/antike/mitarbeiter/menn/contents.

Menn, Stephen. "Plato and the Method of Analysis." *Phronesis* 47, no. 3 (2002): 193–223.

Miller, Andrew M. *Greek Lyric: An Anthology in Translation*. Indianapolis, IN: Hackett, 1996.

Modrak, D. K. "Aisthēsis in the Practical Syllogism." *Philosophical Studies* 30, no. 6 (1976): 379–91.

Moss, Jessica. *Aristotle on the Apparent Good*. Oxford: Oxford University Press, 2012.

Moulton, Samuel T., and Stephen M. Kosslyn. "Imagining Predictions: Mental Imagery as Mental Emulation." *Philosophical Transactions of the Royal Society B: Biological Sciences* 364, no. 1521 (2009): 1273–80.

Mueller, Ian. "Aristotle on Geometrical Objects." *Archiv für Geschichte der Philosophie* 52 (1970): 156–71.

Nagel, Ernest. *The Structure of Science*. Chicago: University of Chicago Press, 1961.

Netz, Reviel. "Aristotle's Metaphysics M3: Realism and the Philosophy of QUA." Stanford/Princeton Working Papers in Classics, 2006.

Netz, Reviel. *The Shaping of Deduction in Greek Mathematics*. Cambridge: Cambridge University Press, 2004.

Nicholson, Nigel. *Aristocracy and Athletics in Archaic and Classical Greece*. Cambridge: Cambridge University Press, 2005.

Nietzsche, Friedrich. *Beyond Good and Evil*. Translated by Judith Norman. Cambridge: Cambridge University Press, 2002.

Nussbaum, Martha C. *Aristotle's De Motu Animalium*. Princeton, NJ: Princeton University Press, 1978.

Nussbaum, Martha C. *The Fragility of Goodness*. Cambridge: Cambridge University Press, 1986.

Nussbaum, Martha C. *The Therapy of Desire*. Princeton, NJ: Princeton University Press, 1994.

O'Connor, David K. "Rewriting the Poets in Plato's Characters." In *The Cambridge Companion to Plato's* Republic, edited by G. R. F. Ferrari, 55–89. Cambridge: Cambridge University Press, 2007.

Pearson, Giles. *Aristotle on Desire*. Cambridge: Cambridge University Press, 2012.

Pink, Thomas, and M. W. F. Stone, eds. *The Will and Human Action: From Antiquity to the Present Day*. London: Routledge, 2014.

Polansky, Ronald. *Aristotle's De Anima*. Cambridge: Cambridge University Press, 2007.

Polansky, Ron, and John Fritz. "Aristotle on Accidental Perception." In *Ar-*

istotle—Contemporary Perspectives on His Thought: On the 2400th Anniversary of Aristotle's Birth, edited by Demetra Sfendoni-Mentzou, 125–50. Berlin, Germany: De Gruyter, 2018.

Porphyry. *On Aristotle's Categories*. Translated by Steven K. Strange. Ithaca, NY: Cornell University Press, 1992.

Price, A. W. "Aristotle, the Stoics, and the Will." In *The Will and Human Action: From Antiquity to the Present Day*, edited by Thomas Pink and M. W. F. Stone, 29–52. Routledge, 2014.

Pylyshyn, Zenon W. "Mental Imagery: In Search of a Theory." *Brain and Behavioral Sciences* 25 (2002): 157–238.

Pylyshyn, Zenon W. "Return of the Mental Image: Are There Really Pictures in the Brain?" *Trends in Cognitive Science* 7, no. 3 (2003): 113–18.

Reeve, C. D. C. *Philosopher-Kings*. Princeton, NJ: Princeton University Press, 1988.

Robinson, Richard. "Up and Down in Plato's Logic." *American Journal of Philology* 84, no. 3 (1963): 300–303.

Rorty, Amélie Oksenberg. "Aristotle on the Metaphysical Status of 'Pathe.'" *Review of Metaphysics* 37, no. 3 (1984): 521–46.

Rorty, Amélie Oksenberg, ed. *Essays on Aristotle's Ethics*. Oakland: University of California Press, 1980.

Rosenmeyer, Thomas G. "Plato's Hypothesis and the Upward Path." *American Journal of Philology* 81, no. 4 (1960): 393–407.

Ross, David. *Aristotle*. 2nd ed. London: Methuen and Company, 1930 [1923].

Scheiter, Krisanna M. "Images, Appearances, and Phantasia in Aristotle." *Phronesis* 5 (2012): 251–78.

Schofield, Malcolm. "Aristotle on the Imagination." In *Articles on Aristotle 4: Psychology and Aesthetics*, edited by Jonathan Barnes, Malcolm Schofield, and Richard Sorabji, 103–32. London: Duckworth, 1979.

Segvic, Heda. *From Protagoras to Aristotle: Essays in Ancient Moral Philosophy*. Princeton, NJ: Princeton University Press, 2008.

Sellars, Wilfrid. "Empiricism and the Philosophy of Mind." In *Minnesota Studies in the Philosophy of Science*, edited by Herbert Feigl and Michael Scriven, 253–329. Minneapolis: University of Minnesota Press, 1956. Published separately as *Empiricism and the Philosophy of Mind*. With an introduction by Richard Rorty and a study guide by Robert Brandom. Cambridge, MA: Harvard University Press, 1997.

Sellars, Wilfrid. *Science, Perception and Reality*. London: Routledge and Kegan Paul, 1963; reissued in 1991 by Ridgeview, Atascadero, CA, 1963.

Sherman, Nancy. *Making a Necessity of Virtue: Aristotle and Kant on Virtue*. Cambridge: Cambridge University Press, 1997.

Silverman, Allan. "Plato on Phantasia." *Classical Antiquity* 10, no. 1 (1991): 123–47.

Sorabji, Richard. *Aristotle on Memory.* 2nd ed. Chicago: University of Chicago Press, 2002.

Sorabji, Richard. "Aristotle on the Rôle of Intellect in Virtue." *Proceedings of the Aristotelian Society*, n.s. 74 (1973–1974): 107–29..

Sorabji, Richard. "The Concept of the Will from Plato to Maximus the Confessor." In *The Will and Human Action: From Antiquity to the Present Day*, edited by Thomas Pink and M. W. F. Stone, 6–25. London: Routledge, 2014.

Sorabji, Richard. *The Philosophy of the Commentators, 200–600 AD: A Source.* (Volume 1: Psychology). London: Duckworth, 2005.

Steiner, Deborah T. *Images in Mind: Statues in Archaic and Classical Greek Literature and Thought.* Princeton, NJ: Princeton University Press, 2001.

Striker, G. "Two Kinds of Deliberation: Aristotle and the Stoics." Unpublished manuscript, 2015.

Tait, W. W. "Noēsis: Plato on Exact Science." In *Reading Natural Philosophy: Essays in the History and Philosophy of Science and Mathematics to Honor Howard Stein on His 70th Birthday*, edited by D. Malament, 11–31. Chicago: Open Court, 2002.

Thompson, Michael. *Life and Action: Elementary Structures of Practice and Practical Thought.* Cambridge, MA: Harvard University Press, 2008.

Thompson, William L., Scott D. Slotnick, Marie S. Burrage, and Stephen M. Kosslyn. "Two Forms of Spatial Imagery: Neuroimaging Evidence." *Psychological Science* 20, no. 10 (2009): 1245–53.

Unguru, Sabetai. "History of Ancient Mathematics—Some Reflections on the State of the Art." *Isis* 70, no. 4 (1979): 555–65.

Unguru, Sabetai. "On the Need to Rewrite the History of Greek Mathematics." *Archive for the History of Exact Sciences* 15, no. 1 (1975): 67–114.

Vernant, Jean-Pierre. *Myth and Thought among the Greeks.* New York: Zone Books, 2006.

Weinman, Michael. *Pleasure in Aristotle's Ethics.* London: Continuum, 2007.

Wiggins, David. "Deliberation and Practical Reason." In *Essays on Aristotle's Ethics*, edited by Amélie Oksenberg Rorty, 221–40. Berkeley: University of California Press, 1980.

Wittgenstein, Ludwig. *Philosophical Investigations.* 3rd ed. German text with revised English translation. Translated by G. E. M. Anscombe. Oxford: Blackwell, 2001.

Yurdin, Joel. "Aristotelian Imagination and the Explanation of Behavior." In *Ancient Perspectives on Aristotle's* De Anima, edited by Gerd Van Riel and Pierre Destrée, 71–88. Leuven: Leuven University Press, 2009.

INDEX

Note: Page references in *italics* refer to figures.

abstraction, 73, 76, 77, 85; Lear's view of, 74–75; Locke's understanding of, 74; Mueller's view of, 73–74; psychological interpretation, 74; semantic interpretation, 75–77, *76*

"accidental/accident" concept, 61, 65–66, 70

acrasia, 140n13

actions: course of action, 89, 104–5, 106; in practical domain, 88; relationship with deliberation, 108; voluntary and involuntary, 111–12

actuality, 59, 77, 80, 83, 86

affective bias model, 56–57

affects, 141n1; modern theories of, 142n23

algebraization, 67

Amphinomus, 102; about geometrical construction, 30; views on geometrical propositions, 29

Analytics (Aristotle), 7

Ancient geometrical practice, 67

animal(s): desire in, 88–89; imagination in, 136n4; imagination role in desire, 103–4; movement, 91–92; passion in, 118; perception in, 89; self-motion, 103

Anscombe, G. E. M., 107

appearance(s), 15, 19, 36, 41, 137n9; phantasia, 44; phantasma, 44, 48; Plato's views about, 15–16; power of, 7; psychological, 7; and shame, 116. *See also* imagination

appetite, 89, 93, 94, 103

Aristippus, 20

aristocracy, 11

Aristotle: actualities, discussion of, 83–84; affective bias model, 56–57; analogy between geometrical analysis and deliberation, 102; analogy between theory and practice, 88; argument on doctrine of intermediates, 135n53; argument on mathematical objects, 63–64; comparison of deliberation to geometrical analysis, 101–2; comparison of imagination to light, 53–54; conception of wish, 93–94, 96, 97–98; concern about geometrical inference, 70–72, *71*; criterion for psychological division, 33–35; deliberative imagination, 102–5; deliberative rhetoric characterization, 141n16; denying identification of perception, 27; dream, definition of, 57–58, 59; about geometrical construction, 30, 35–36; geometrical practice, understanding of, 64–67; geometry,

characterization of, 7–8, 25, 79–80; identifying formal cause of animal movement, 91–92; illusions, descriptions of, 55; imagination, characterization of, 18, 36–37, 52, 61–62; interpretation of psychology of desire, 97; logical and cognitive points of science, 65; objection to realism and relativism, 90, 97; "other animals," description of, 140n9; passions, descriptions of, 118; perceptible shape, definition of, 21; phantasia, definitions of, 41–46; phantasma, characterization of, 58; Plato on mathematical views, attribution to, 23–24; Platonic phantasia, adoption of, 37–38; practical reasoning *vs.* theoretical reasoning, 106–8; principle of unchanging things, 29; psychological theory, 17–18; quality, division between four kinds of, 119–20; about rational desire, 90–91, 92; refuting identification of perception, 50–51; shame, descriptions of, 113–14, 117–18, 121–22; sun scenario adopted from Heraclitus, 55–56; universal logical form of geometrical demonstrations, 72–73; views on mathematical objects, 47–48; wartime scenario, description of, 106
astronomy, 24, 64, 79, 135n56

Barnes, Jonathan, 82
belief, 38, 41–42, 51–52, 56, 96, 108, 137n21
Benardete, Seth, 55
blameworthiness, 111–12, 123
blushing: due to psychological passion, 120–21; constitutional, 120
Bolton, Robert, 53, 58–59
Bulmer-Thomas, Ivor, 135n56
Burnyeat, Myles F., 23

calculation, 20, 75, 79, 89, 102–5; practical, 21
calculative imagination, 102–3
Caston, Victor, 53
Categories (Aristotle), 119
Cherniss, Harold, 28
choice, 89, 94, 99, 104–9, 111, 122. *See also* deliberation
co-exact attributes, 70, 72, 77
cognition: geometrical, 83–84, 86, 94; power of, 125, 127–28; presence of object of, 49, 50
collection method, 33, 78
Collection (Pappus), 68
Commentary on the First Book of Euclid's Elements (Proclus), 8
conditional realism, 64
conscious stimuli, 58
constructivists, formalists *vs.*, 30–31
conviction, 41, 136n1
copying, 16
Corkum, Phil, 64
Cornford, Francis MacDonald, 22
courage, kinds of, 112
Cross, R. C., 23
cultic contemplation, 10

David (sculpture by Michelangelo), 16–17, 18
dead in Greek literature, 9–10
De Anima (Aristotle), 7, 33, 37, 40, 46, 49, 102, 105, 117, 137n9
deliberation, 94, 99; and acting on one's choice, 108–9; Aristotle's analogy with geometrical analysis, 102; Aristotle's comparison to geometrical analysis, 101–2; hypothetical or provisional nature,

100–101; instrumental and deductive interpretations, 99–100; modal and temporal constraints, 108; *vs.* practical syllogism, 107; progression of geometrical "actions," 100; relationship with action, 108. *See also* imagination
deliberative imagination, 102; as cognitive function, 105–6; courses of action, 104–5; *vs.* perceptual imagination, 102–4
De Memoria (Aristotle), 84
Democritus, 18
Descartes, René, 142n14
desire, 141n6; absent object of, 103; animal movement, 91–92; in animals, 88–89, 94; correctness of, 96; of excellent individual, 96–97; imagination role in, 103–4; in inanimate objects and animals, 91; rational form of, 89–90; relation between form and matter, 91
Dewey, John, 5
diagrams, 29, 30, 47, 72, 75–77, 78–84; importance in mathematics, 78; reasoning by, 84; role in geometry, 66–67; use in *reductio* proofs, 70
Dillon, John, 25
discerning functions of imagination, 54–55
discursive reasoning, 36
discursive thinking, 20, 22, 24, 25, 33, 51, 60, 93, 109
disrepute (*adoxia*), 114, 116
disturbance, 113, 117
division method, 4–5, 33, 79, 113; access to geometrical objects, 35–36; phantasia from belief, 41–42, 46; phantasia from perception, 40–41, 46; psychological division, 33–35
"drawing beyond the side" method, 138n29
dreams: Aristotle's definition of, 57–59; falsity of, 137n21

Elements (Euclid), 31
emotions, 56, 119–20, 141n1, 142n14
Empedocles, 12, 49, 50
emphasis, 18
consciousness, 10, 45, 126, 127
endoxa, 37, 40
energeia as "actuality," 80
Epictetus, 125
Ethics (Aristotle), 33–34, 53, 67, 69, 102, 104, 105
Euclid, 22, 31–32, *32*, 139n30
Eudemian Ethics (Aristotle), 29, 113, 114–15
Eudemus the Peripatetic, 138n29
Eudoxus, 30
Euthydemus (Socrates), 78
exact attributes, 70, 77
"expansion," doxastic, 69–70

faint noise during sleep, 58
"figures" concept, 22–23
formalists *vs.* constructivists, 30–31
Fortenbaugh, William W., 117–18
Frede, D.: discerning functions of imagination, 54–55; about imagination, 52–53
freedom, 4, 94
free will, 4
functional grouping, 4, 7
funerary sculptures, 11

Galton, Francis, 124; description of property of imagination, 126; explanation of mental images, 127
Garland, Robert, 10
Geminus, 138–39n29
generality, 65, 71, 77

geometrical analysis: Aristotle's analogy of deliberation with, 102; Aristotle's comparison of deliberation to, 101–2
geometrical cognition, 47, 80, 83, 94; imagination role in, 84; stages, 86
geometrical constructions, 79; Aristotle about, 30; Euclid's illustration of, 31–32, *32;* Plato about, 30
geometrical demonstration, 27
geometrical inference, 66–67, 70–72, *71*
geometrical objects, 27, 28; "alienated" from perceptible representation, 47; cognitive status of, 87; kinesthetic construction of, 46
geometrical propositions, 29
geometrical thinking: Aristotle's conception of geometry, 79–80; geometrical cognition, 84, 86; Heath's interpretation of geometry, 80–82, *81*, *82*; perception, 85; Plato's views on mathematics, 78; "quasi-perceptual intuition" of Aristotle, 83; reasoning by diagrams, 84; universality of geometrical judgment, 86–87
geometry, 20, 79, 84; Aristotle's conception of, 25, 65, 67, 79–80, 82; axiomatic structure to, 36; cognitive status of geometrical objects, 87; constructive process of alteration in, 28; Euclidean codification of, 22; Heath's interpretation of, 80–81, *81*, *82*; intellectual forms, 85–86; paradox of, 28, 31; Plato's conception of, 25; role of diagrams in, 66–67; use of potentiality in, 83
Glaucon, 14, 25, 28
Gombrich, E. H., 11
Grimaldi, William M. A., 113–15

Hackforth, Reginald, 23
Handbook (Albinus), 37
Havelock, Eric A., 14
Heath, Thomas, 80–82, *81*, 139n29
Heraclitus, 21, 55
Herodotus, 12, 20
Hippocrates of Chios, 70
Hobbes, Thomas, 10, 125
Homer, 13, 14, 37, 50, 112
human forms, 10–11
Humean theory of practical reasoning, 93
hypothesis concept, 22

"iffy" facts, 6
ignorance, kinds of, 141n2
illusion: Aristotle's descriptions of, 55–56; bias model of, 57
image(s): fixative function of, 12; likeness, 9, 12; logic of replacement, 12; mimetic function of, 11; Plato's views about, 16; political function of, 12; representation in Greek sources, 9–10; Socrates' definition of, 14–15; types of concepts, 10–11; and Xenophanes of Colophon's views, 12–13
imagination, 5, 17, 36–37, 43, 124; an innovation of theoretical philosophy, 48; Aristotle's comparison to light, 53–54; Aristotle's definition of, 52, 61–62; in Aristotle's descriptions of illusions, 55; association with perception, 44–45; augmentation of faint perceptions in sleep, 58; deliberative, 102–3; discerning

functions, 54–55; and dreaming, 59; "echo" interpretation, 54; Frede about, 52–53; freedom from determination, 94–95; Galton's description of, 126–27; iconoclasts's views on, 45, 46; iconophiles's views on Aristotle's descriptions, 45–46; noncontact and, 51; "parasitic" understanding of, 58, 60; and phantasia, 40–44; in philosophical modernity, 125–26; power of cognition, 127–28; powers of judgment, 51–52; relationship to intellect, 126; "reproductive" conception, 46; role in desire, 103–4; role in geometrical cognition, 84; as special function of perception, 51; in theoretical domain, 88; theory of, 18, 26; variation in people's understandings, 128–29. *See also* perception
imitative art, 14–15; comparative freedom of, 15; types of, 16
inanimate objects and animals, 91
indwelling imagination, 102–3
intellect, 34, 51, 83, 84, 96; discursive thinking and, 22, 24, 32; noemata, 61; noesis, 80; using phantasmata, 43; relationship of imagination to, 46–47, 55, 97, 126
intelligible figures, reasoning about, 20, 22–26
intelligible matter, 63, 82
intermediacy, doctrine of, 25
intermediates, 23–25, 31
Irwin, Terence H., 91

Johansen, Thomas Kjeller, 34, 60
Jonas, Hans, 104
judgment, 38, 40, 102, 106, 123; belief, 41–42, 43, 46; powers of, 51–52; universal 86

Kant, Immanuel, 3, 128; empirical consciousness, 126; practical rule, 93; theory of practical reasoning, 93
knowable object, 21
Konstan, David, 114, 115

Lachterman, David Rapport, 29
Lakatos, Imre, 4
Lännström, Anna, 96
Lear, Jonathan, 34; about mathematics, 64; view of Aristotelian abstraction, 74–75
legislation, art of, 111
Leighton, Stephen R., 116–17
Leodamas, 70
Locke, John, 74–75
love, 112, 115

Makin, Stephen: claims to Aristotle's "quasi-perceptual intuition," 83; Heath's interpretation of geometry, 80–82, *81*, *82*
Manders, Kenneth, 70
"manifest image" of man, 3, 6, 7
materialist theory of sense perception, 18
mathematical genesis, 30
mathematical objects, 21, 23–25, 46–47, 78, 85, 125; Aristotle's arguments on, 63–64; instantiated in physical object, 75; ontological debate concerning, 65
mathematics, 29–30, 64, 83; conversion in, 69; perceptible objects in, 85; Plato's views, 78–79. *See also* geometry
McDowell, John, 100
memory, 36; dependence on phantasmata of time, 61; of intelligible things, 60, 61
Menaechmus, 30, 66

Menn, Stephen, 141n17
Meno (Plato), 20–21
mental imagery/images, 17, 126–27
mentality, 3–4, 5
Metaphysics (Aristotle), 7, 63, 64, 79, 82, 83, 119
Michelangelo, 16–17, 18
mindedness, 3, 4, 127
moderation in shame, 117
Moss, Jessica: about Aristotle's comparison of memory, 60; cognition or discernment, 55; description of bad, 92; about imaginings, 59
Mueller, Ian, 64, 73
multi-layered universe, 25
multiplicity, 25, 90

Nagel, Ernest, 5
natural philosophy, 63
Netz, Reviel, 64
Nicholson, Nigel, 11
Nicomachean Ethics (Aristotle), 113
Nietzsche, Friedrich, 3
noesis, 80
noncontact, 51
Nussbaum, Martha C., 44, 115

Occam's razor, 5
O'Connor, David K., 37
On Nature (Empedocles), 49
On the Heavens (Aristotle), 29
On the Motion of Animals (Aristotle), 46
On Truth (Protagoras), 18
opinable object, 21–22
"ordered" problem, 102, 104

Pappus of Alexandria, 68–69
Parva Naturalia (Aristotle), 7, 61
passions, 93, 96, 110, 111, 113, 141n1; in animal, 118; Aristotle's descriptions of, 117–18, 121; cognitive aspect, 115; to psyche and body, 120–21
pathos, 119, 120
perceptible figures, reasoning about, 20–21, 23–25
perceptible objects, 25, 48, 85
perceptible things/matter, 3, 21, 24, 25, 63, 85
perception, 19, 43, 50, 54, 85, 124; of animals, 50–51, 89; association of imagination with, 44–45; dependency of, 50; from geometrical point of view, 65; of geometer's knowledge on things, 79; identification of, 27; imagination as special function of, 51; phantasia division from, 40–41; powers of, 52. *See also* imagination
perceptual imagination, 102–3
Phaedrus (Plato), 33
phainomenon, 15
phantasia, 7, 37–39, 40, 60, 125, 137n9; arguments for individuation through division, 42–43; Aristotle's definitions of, 43–44; Benardete about, 55; Caston about, 53; and definition of shame, 115–16; derivation from "light", 53–54; descriptions of, 40; division from belief, 41–42; division from perception, 40–41; from dreams, 137n21
Philebus (Plato), 17
physicalistic theory of mind 49–50
Plato, 4, 112; analysis method, 70; appearance, views about, 15–16, 48; argument on self-refuting, 134n37; calling art as *phantastike*, 124–25; critique of imitative art, 14; critique of Protagoras' "art of

politics" concept, 19; deception concept, 7; denial of possibility of knowledge of sensible things, 21; denying identification of perception, 27; discussions of images and appearances, 18; dualistic view of diagrammatic discovery and philosophical knowing, 84–85; explanation of derivation of image, 15; about geometrical construction, 30; about geometry studies, 32–33; images of virtue, views about, 16; intermediacy, doctrine of, 25; mathematics, views on, 78; methodological comments on astronomy, 135n56; about natural sciences, 25–26; observations on geometrical objects, 28; paradox of geometry, 31, 37; phantasia, 37–39; powers of psyche, 22; psychological theory of mental imagery in *Philebus*, 17; reasoning about intelligible figures, 20, 22–26; reasoning about perceptible figures, 20–21, 23–25; usage of division and collection methods, 33

Poetics (Aristotle), 43

Polansky, Ron, 55, 141n6

Porphyry, 119

Posterior Analytics (Aristotle), 69, 71, 72, 82

potentiality in geometry, 79–80

practical reasoning, 93, 98, 106–7, 108

practical wisdom, 100, 106

praiseworthiness, 122–23, 142–43n26

Proclus, 8, 22, 24; claims of Plato's analysis method, 70; geometrical proposition, 31; report on division method, 33; views on geometrical propositions, 29, 31–32

Protagoras (Plato), 112–13

Protagoras, 18; art of politics, 18–19; Plato's critique, 19; sophistical refutation of geometers, 19–20

psyche: body and, 120–21

psyche, powers of, 3–5, 17, 22, 27; calculative and perceptual imagination, 102–3; division method, 33–35; geometrical construction, 28–33; imagination, 36–39

psychological appearances, 7, 119, 127

psychological theory, 5, 17–18, 124, 127; empirical, 126; of mental imagery, 17; negative heuristic of, 4

quality, kinds of, 119–20

rational deliberation, 111

rational powers, 108–9, 141n17

rational psyche, 49

reality, 9, 15, 19, 64; external, 10; independent, 18; of supersensory object, 5

real psychological ascriptions, 7

Reeve, C. D. C., 22–23

relativism, 19, 97

Republic (Plato), 14, 21–23, 37–38

Rhetoric (Aristotle), 108, 113, 117, 142n12

Rival Lovers (Plato), 78

Rorty, Amélie, 119

Ross, David, 51; about dream, 59; about Humean interpretation on imagination, 58

Schofield, Malcolm, 37, 54; discerning functions of imagination, 54; views on Aristotelian imagination, 44

Segvic, Heda, 101

Sellars, Wilfrid, 3, 6
sensory perception, 49
shame, 110–11, 142n12; from act of imagining one's disrepute, 116–17; Aristotle's description of, 113–14, 117–18, 121–22; avoidance of, 112, 116, 122; and blameworthiness, 111–12; capacity to experience, 123; disrepute and, 114; doxastic commitments to others, 114–15; expressions of, 142n23; moderation in, 117; paradigmatic bodily state of, 119; phantasia and, 115–16; in Platonic myth, 113; political function of, 112–13; and praiseworthiness, 122–23; rational and political aspect of, 114–15; sociality of, 112
Sherman, Nancy, 115
Silverman, Allan, 37
Socrates, 37; constructive process of alteration in geometry, 28; definition of shape, 20–21; about discursive thinking, 25; explanation of images derived from appearances, 14–15; on mathematical objects, 78
Sophist (Plato), 16, 42, 136n1
Sophistical Refutations (Aristotle), 67, 69
Sorabji, Richard, 44
Speusippus, 29; derivation of universe, 25; about geometrical construction, 30
statues, 10–11
Steiner, Deborah T., 12
Striker, Gisela, 106, 140n15
subconscious stimuli, 58
subtraction, 29, 73. *See also* abstraction
sun scenario, 55–56

Tait, W. W., 21
temper, 89, 92–94, 115
Theaetetus (Plato), 19
theoretical reasoning, 106–7, 108
theory change, 4–5
touch, sense of, 55, 56, 104, 136n4

Unguru, Sabetai, 67, 138n7
universality of geometrical judgment, 86–87
utility concept, 18–19

value realists *vs.* value relativists, 89–90
Vernant, Jean-Pierre, 10–11
visible things, 16

Weinman, Michael, 34
Wiggins, David, 99–100
wish(ing), 89; Aristotle's conception of, 93–94, 96, 97–98; coincides with supposition, 92; good life, 97, 100, 101; rational desire, 90–91, 92; realism *vs.* relativism, 89–90; role in diversity of communities, 95; of statesman, 95
Woozley, A. D., 23

Xenophanes of Colophon: against anthropocentrism of representational art, 13; against anthropomorphized conception of gods, 12–14

Zeno the Stoic, 125